Environmental Philosophy

Environmental Philosophy
A Collection of Readings

edited by

Robert Elliot
and
Arran Gare

The Pennsylvania State University Press
University Park and London

Published in the United States of America by
The Pennsylvania State University Press

Manufactured in Australia

Library of Congress Cataloging in Publication Data

Main entry under title:

Environmental philosophy.

  Bibliography: p.
  Contents: Ethical principals for environmental
protection / Robert Goodin — Political representation
for future generations / Gregory S. Kavka and Virginia
L. Warren — On the survival of humanity / Jan
Narveson — [etc.]
  1. Human ecology — Moral and ethical aspects —
Addresses, essays, lectures. 2. Environmental protection —
Moral and ethical aspects — Addresses, essays,
lectures. 3. Human ecology — Philosophy — Addresses,
essays, lectures. I. Elliot, Robert, 1950-   .
II. Gare, Arran, 1948-   .
GF80.E59   1983   304.2   82-21745
ISBN 0-271-00355-3
ISBN 0-271-00356-1 (pbk.)

Contents

Acknowledgments

We would like to thank authors and publishers for permission to reproduce the following copyright works:

Man's Responsibility for Nature by John Passmore, published by Gerald Duckworth & Co Ltd (1974); *A Sand County Almanac* by Aldo Leopold, published by Oxford University Press (1949 and 1966); *The Moral Status of Animals* by Stephen R.L. Clark, published by Oxford University Press (1977); *Keepers of the Game: Indian-Animal Relationships and the Fur Trade* by Calvin Martin, published by the University of California Press (1978); "No Moral Nukes", *Ethics* 90 (April 1980) by Robert Goodin, published by the University of Chicago Press; and "Duties Concerning Islands", *Encounter* (1982) by Mary Midgley.

Contributors

Robin Attfield, Department of Philosophy, University College, Cardiff, Wales.

J. Baird Callicott, Department of Philosophy, University of Wisconsin at Stevens Point, Stevens Point, Wisconsin, U.S.A.

Stephen R.L. Clark, Department of Moral Philosophy, University of Glasgow, Glasgow, Scotland.

Robert Elliot, Department of Education Studies. Brisbane College of Advanced Education, Kelvin Grove Campus, Brisbane, Australia.

Arran Gare, formerly of the Department of Philosophy, University of Queensland, Brisbane, Australia.

Robert Goodin, Department of Government, University of Essex, Colchester, England.

C.A. Hooker, Department of Philosophy, University of Newcastle, Newcastle, Australia.

Gregory S. Kavka, Department of Philosophy, University of California at Irvine, Irvine, California, U.S.A.

Mary Midgley, formerly of the Department of Philosophy, University of of Newcastle, Newcastle-on-Tyne, England.

Jan Narveson, Department of Philosophy, University of Waterloo, Waterloo, Canada.

Richard Routley, Department of Philosophy, Research School of Social Sciences, Australian National University, Canberra, Australia.

Janna Thompson, Department of Philosophy, La Trobe University, Melbourne, Australia.

Mary Anne Warren, Department of Philosophy, San Francisco State University, San Francisco, California, U.S.A.

Virginia Warren, Department of Philosophy, Chapman College, Orange, California, U.S.A.

Introduction

ROBERT ELLIOT and ARRAN GARE

There are many reasons for thinking that humankind's relationship with the natural environment needs to be reassessed. Population growth and economic expansion are eroding the resource base upon which human civilization depends. Widespread pollution of the land, the sea and the air detracts from the quality of human life. The development of nuclear technology imposes grave risks to present and future people. The destruction of species and the despoliation of wilderness threaten to destabilize ecosystems that are necessary to the health of the biosphere and, therefore, to human life.

Concern for present and future human beings should be sufficient motivation for calling into question our relationship with the natural environment and modifying our attitudes to it. There are other reasons as well. Our policies threaten nonhuman animal populations. This may happen when, for example, valleys are flooded to provide hydro-electric power. In such cases animals are killed and others are compelled to leave established habitats. The annual, systematic butchery of Candian harp seals and the slaughter of elephants and whales are more obvious examples of human policies that adversely affect animals. There is some plausibility in the suggestion that any morality which permits such acts, expecially where they are not necessary for securing human welfare, is warped.

Certain policies may also be condemned not just because they harm human beings or nonhuman animals but because they threaten parts of the natural environment. Policies, such as the inundation of Lake Pedder in Tasmania, that involve the destruction of wilderness may be condemned because they destroy what has intrinsic value. In other words they are condemned not just because they destroy the objects of certain recreational interests but because they destroy things which are valued in themselves and independently of uses to which they might be put.

An environmental ethic may be thought of as a systematic ethic which includes the three distinct elements mentioned so far. It is an ethic which allows that future generations, nonhuman animals and nonsentient nature are all morally considerable. They may not be counted considerable in exactly the same ways. For example, the claim that trees have moral rights might be resisted for theoretical reasons. However, the ethic might allow that trees do count in moral calculations because they have intrinsic value. What is significant is that an environmental ethic represents a decided shift from traditional ethics which place human beings at the centre of the moral universe to an ethic of much wider scope. Some of the papers collected in this volume are efforts towards developing the detail of, and justifying, this expanded ethic.

The papers in part 1 by Robert Goodin, Gregory S. Kavka and Virginia Warren, Jan Narveson, C. A. Hooker and Janna Thompson, are pertinent to any attempt to understand the significance of future generations in morally assessing environmental policy. Robert Goodin, in his "Ethical Principles for Environmental Protection", stresses the need for a moral approach to policy assessment. He notes that the ethical principles that should constrain policy need to be simple and easily applicable in real-life situations. Goodin begins with criticisms of a widely accepted principle which urges the maximization of expected benefits. He then reviews six alternatives to this principle that he thinks provide a better moral basis for environmental policy making.

Gregory S. Kavka and Virginia Warren, in their "Political Representation For Future Generations", address the problem of how future people might be represented in a democratic political system. They establish that future people *can* be represented, go on to argue that they *should* be represented, and make some suggestions about how this might be achieved.

Jan Narveson, in his "On the Survival of Humankind", asks whether there is a direct obligation to ensure the survival of the human species. He claims that there is not and that there is no intrinsic value in extending human occupancy of the universe into the future. However, he does urge that, if there are going to be people around in the future, we have obligations to them and should consider them when we formulate our policies. Narveson then inquires into whether or not the obligations we do have could necessitate a eugenics programme. While conceding the theoretical possibility that it does, he goes on to claim that in any likely state of affairs a eugenics programme would violate, unjustifiably, important moral principles.

C. A. Hooker's contribution, "On Deep Versus Shallow Theories of Environmental Pollution", begins with a statement of the role philosophy has to play in resolving environmental problems. He suggests that "the philosopher's role is much more than auxiliary analyst, it involves the articulation of a wise way of life and the diagnosis of the deep roots of unwisdom in extant cultures". Hooker spends some time giving a critical account of Passmore's theory of pollution in order to show the importance of holistic, rather than piecemeal, examination of the problem and solutions to it. He stresses that many preferred solutions to problems of pollution are bound by the same structures which gave rise to the problems in the first place. What adequate solutions require is a systems-theoretic approach coupled with a desire for flexibility of policy and a vision of a "healthy society".

Janna Thompson, in her "Preservation of Wilderness and the Good Life", presents a position which reduces the conflict between shallow and deep, long-range environmentalism. In an important paper, "The Shallow and the Deep, Long-Range Ecology Movement", published in 1972, Arne Naess distinguished between environmentalists who base their conservationist and preservationist policies on considerations to do with human welfare and interests, and those who suggest that all of nature has intrinsic value. Naess pointed out that while there may be some concurrence on matters of policy between these two streams of environmentalism, there will nevertheless be some policy matters on which they disagree. There may be conflicts between human interests, the "interests" of wild animals and the "interests" of tracts of wilderness, which are irresolvable. To the extent that there are such policy disagreements, radical environmentalists will find it difficult to win acceptance for their policies. However, Thompson suggests that human self-realization and happiness can only be fully achieved when natural systems and living creatures are respected in their own right and not treated merely as instruments or as objects of dominance. Thompson argues that the search for a more liberated, ideal social existence will likely bring with it changes in attitudes towards the natural environment and reduce the conflicts between human beings and their natural environment. Thompson stresses that moral revolutions go hand in hand with social revolutions. She goes on to make positive suggestions about social change that would make an environmental ethic workable and widely acceptable.

The papers by Mary Anne Warren, Holmes Rolston III, Mary Midgley and Stephen R. L. Clark deal with rather different issues to those dealt with in part 1. They deal specifically with questions about the moral significance of nonhuman animals and the natural environment.

Mary Anne Warren's contribution, "The Rights of the Nonhuman World", takes up the conflict between the views of animal liberationists such as Peter Singer and Tom Regan and those of Aldo Leopold. Leopold's environmental ethic is said to be holistic and genuinely new. The ethics propounded by Singer and Regan are said to be individualistic and to be merely extensions of existing ethical systems. Warren suggests that the two views are better thought of as complementary, rather than competing, and that *both* are required in order to accommodate all of the relevant moral considerations.

Holmes Rolston III, in his "Are Values in Nature Subjective or Objective?", attacks the subjectivization of values, and claims it is one of the roots of the environmental crisis. While accepting that experience is mediated by our cultural "education", Rolston argues that the values we experience are *in the world* and that we, as valuers, are part of this world. He then develops the notion of the experience of value as relational, and clarifies the connection between experience and its objective base by considering a number of rival accounts of value. He concludes that nature is the bearer of a value which "both constrains and enables the role we humans are called to play".

Mary Midgley, in "Duties Concerning Islands", considers the question of whether Robinson Crusoe had any duties; whether there would have been anything wrong with Crusoe devastating his island as he was about to leave it. Midgley believes that a serious consideration of this imaginary situation will yield important results concerning the basis and extent of an environmental ethic. She is critical of the dominant, individualistic ethical theories, which seem to imply that there is nothing wrong with Crusoe's action, and which present themselves as superior to our intuitions. She then argues for a notion of duties that transcends the symmetrical relationships involved in contract theory.

Stephen Clark, in "Gaia and the Forms of Life", has attempted to elaborate and defend an holistic environmentalist ethic on the basis of the notion of Gaia. Emphasizing the complex self-regulating nature of the biosphere, Clark argues that the widespread desire to view the beauties of nature reflects a deep-rooted recognition of our dependence on a healthy, functioning environment, and that this environment can only be considered healthy when it functions without our intervention. He then attacks the subjectivism and individualism of modern ethics in favour of an attitude in which we would acknowledge our recognition of ourselves as merely a part of a dynamic whole.

The papers in part 3 by Robin Attfield, J. Baird Callicot and Richard Routley take up a different set of issues again. They are best understood in the light of the first modern philosophical work that extensively dealt with humankind's relationship with nature, namely, John Pass-more's *Man's Responsibility for Nature,* which was published in 1974. Passmore set out to discover whether there are sound arguments for reducing pollution, husbanding resources for future generations, minimizing population growth and preserving species and wilderness. Passmore concluded that there are, and that the roots of such arguments are to be found in traditional Western religious and ethical systems. Much of Passmore's work traces attitudinal themes concerning the natural environment, in order to discover whether there are aspects of our cultural tradition that can be built on in order to solve environmental problems. He concludes that there are, and that they may be developed within the context of liberal, democratic social arrangements.

Other philosophers have been more critical of Passmore's favoured traditions. They have argued that Western ethical traditions are deficient in providing for the welfare of future generations, that they are unjustifiably anthropocentric and lacking in concern for nonhuman animals and wild nature. Western traditions have been, in this respect, compared unfavourably with the traditions of other cultural groups such as American Indians and Australian Aborigines. These philosophers have typically called for a radical transformation in our ways of thinking about nature and our relationship with it. Another line of criticism suggests that Passmore's survey and exegesis of Western traditions is deficient, and that these traditions do provide the basis for a much more radical environmental ethic than Passmore believes possible. Both of these lines of criticism are represented in part 3.

Attfield argues that "proposals for an environmental ethic . . . are a re-expression of long-standing themes of Western culture". Like Passmore, Attfield thinks that it is only by exploiting themes already present in Western cultural, ethical and religious traditions that an adequate and useful environmental ethic will be developed. However, Attfield identifies themes that Passmore had not noticed and that do support a radical environmental ethic. This means that Attfield, unlike many of Passmore's critics, does not think that the establishment of an environmental ethic requires anything like a revolution in Western moral thought, since the basis for such an ethic is already present.

Attfield's claims are indirectly called into question by J. Baird Callicott in his "Traditional American Indian and Traditional Western European Attitudes towards Nature". Callicott claims that "American Indian cultures provided their members with an environmental ethical ideal". He goes on to argue that this environmental ethical ideal is superior to any ideal implicit in the dominant Western traditions, as far as promoting a balanced relationship between human beings and their natural environment is concerned.

The same issue is taken up by Richard Routley in "Roles and Limits of Paradigms in Environmental Thought and Action". Routley considers the claim that an adequate solution to environmental problems requires a reorientation of Western attitudes to the natural environment; that is, a shift in the world views, paradigms and such like that are characteristic of Western thought. Routley argues that since there is little chance that such changes will be achieved, environmentalists would do best to argue their case in the context of existing attitudes. He concludes with a discussion of some strategies which could be used to this end.

Part One
Environmental Policy and Human Welfare

Ethical Principles for Environmental Protection*

ROBERT E. GOODIN

For purposes of public policy making, subtle ethical doctrines are invariably translated into simple-minded principles. There are many reasons for this. Some seem to be inherent in the nature of public bureaucracies. Others derive from the nature of moral principles themselves.[1] But whatever its source, this tendency does exist and it powerfully constrains moral analysts of public issues: if their ethical advice is ever to be implemented, it must ultimately be reducible to some such rules of thumb, which can be stated simply yet applied widely. Ethical analysts of public policy, just as more ordinary policy analysts, must conduct and report their research in such ways that policy makers find it "usable".[2]

Here I shall canvass several such simple principles that might be applied in environmental, natural resource and, most especially, energy policy making. One of the great advantages of focusing on specific policy issues in this way lies in the possibility of discovering important limits of and alternatives to rules of thumb, which look compellingly attractive in the general case. After surveying the orthodox "utilitarian/cost-benefit" rule and its shortcomings as a principle for environmental policy making, I shall survey six alternatives to that principle which might be better suited to this particular context.

I
Maximizing Expected Utility

The orthodox rule used in making public decisions generally and environmental ones especially is an updated version of utilitarianism.

* Material from Robert E. Goodin's "No Moral Nukes", *Ethics* 90 (April 1980): 417–49 is reproduced by permission of The University of Chicago Press.

Each alternative course of action is evaluated according to the ratio of its costs to its benefits, rendered commensurable most commonly through monetary equivalents. The option which "maximizes happiness" in the modern sense of having the highest cost-benefit ratio is recommended. Where costs and benefits of each option are probable rather than certain, their expected values (or certainty equivalents) are calculated by discounting each possible cost or benefit by the probability associated with its occurrence. Those who are indifferent to running risks decide strictly on the basis of these expected values, while those who either like or loath risks adjust these figures upwards or downwards, depending on the size of the risks and the intensity of their feelings about them. Such techniques are now widely used in deciding whether and where to pollute the air and water, build cities and dams and airports and power stations. The techniques are familiar, as are the objections to them.[3] Hence my own discussion of them can be rather perfunctory.

The first objection to "utilitarian/cost-benefit/expected-utility" rules is that they are based on individual preferences, which may provide infirm foundations for policy making in various ways. When stating their preferences, people always act at least partially in ignorance. Even if they "knew" all the relevant facts intellectually, they would often be psychologically incapable of conjuring up the sort of vivid image of what it would be like to *experience* that state of affairs which is necessary in order to form a proper preference for it, one way or the other. Furthermore, people's preferences do not predate experience but rather grow out of it. We do not have a very clear idea of our preferences for things lying very far outside either our past experiences or our present possibilities. Hare's proposal for a "trial-design" method of environmental planning — asking people to choose between alternative plans rather than just to state their preferences in the abstract — might go some way towards expanding people's vision, but it cannot go far enough to meet the real objection. The point remains that desires are powerfully adaptive, tailoring themselves to people's histories and possibilities; and if what people want is largely determined by what they get, then it is ludicrously circular and irrational to decide what they should get by asking what they want. My own view is that preferences and goods do come bundled together, but that we can still sensibly determine which bundle leaves people most satisfied. We can argue in that way, however, only if we accept "cardinal" notions of utility, which are an anathema to most economists.[4]

A second objection centres around the commensurability of values which this utilitarian principle presupposes. In one aspect, this is the famous problem of comparing the preferences, desires and utilities of different individuals. Even if we had no qualms about basing policy on people's preferences, we would still need to aggregate their conflicting demands into a single social decision; although there are various ingenious techniques for such interpersonal comparisons, I am persuaded that none can provide an objective basis, and such comparisons must rest instead on ethical postulates.[5]

Another aspect of the commensurability problem, which is more central to the practice of cost-benefit/expected-utility analysis, is that not all goods are tradable for one another or able to be converted into monetary equivalents. The problem is partly with the instruments used to establish equivalence. It is, for example, absurd to suppose that the amount of insurance coverage carried on a life or a historic church represents its full value to society.[6] More fundamentally, the problem is that some things may not be tradable for one another. The stock philosophical example asks how many sweets it would take to induce you to kill your grandmother: the answer, presumably, is that you would not be willing to kill her, no matter how many sweets you were offered; and the reason is not just that you would be satiated before you ate them all but is, rather, that granny ranks lexicographically prior to sweets in your value system.[7] Similarly in environmentalist debates, old-line conservationists often argue that preserving a threatened species or wilderness should rank absolutely prior to economic growth. If there are any such commodities that are not cashable in terms of a common metric, then the summing up implicit in utilitarian/cost-benefit/expected-utility procedures is strictly impossible. It would be like adding apples and oranges.

The third and most common objection to utilitarianism, which also applies to its modern embodiments, is that it is impervious to distributions of happiness. Suppose utility could be maximized by giving everything to one person (a "super-efficient pleasure machine"), while leaving everyone else to starve. Then utilitarianism and its heirs recommend doing so, whereas our very strong moral intuitions dictate otherwise. Cost-benefit analysis has equally distressing implications: it would be better to dam up a river so as to displace a thousand families living in £20,000 houses rather than inconvenience one family living on a £21,000,000 estate.[8] That, of course, assumes crucially that all we care about is the value of the property flooded. We may, however, care not only how high the costs are but also who has to pay them. There is nothing in the logic of expected-utility calculations that forces us to take distributions into account, but neither is there anything that precludes us from doing so. If decision makers care about the distributive impact of their policies, they can — and there is some evidence that they do — build "distributive weights" into their assessments of policy options, boosting the expected utility of policies that allocate benefits to certain favoured groups (e.g., the least advantaged).[9] Still, I take it to be a serious criticism of expected-utility procedures that they treat distributive adjustments merely as an option, whereas on most other principles they are morally obligatory.

This discussion is little more than a sketchy reminder of familiar problems associated with utilitarianism and, *pari passu,* with its modern instantiations, cost-benefit/expected-utility analysis. While these objections are perfectly generaly, they apply with particular force, given some of the peculiar features of environmental decisions. Taken together, they should suffice to motivate the search for alternative principles which can be used to guide decision making on environmental and analogous issues.

II
Keeping the Options Open (Reversibility)

Some of the most poignant pleas for environmental protection are couched in "forever more" terms. We are urged to prevent the *extinction* of certain species, or to prevent the destruction of *irreplaceable* historical landmarks or natural vistas. Many factors might contribute to the power of these appeals, but surely one of the more important is that they violate Arthur C. Clarke's rule, "Do not commit the irrevocable." The Study Group on Critical Environmental Problems explicitly defined its goal as being "to prevent irreversible global damage" and, along these lines, defined as among the most critical pollutants heavy metals such as mercury, which is "a nearly permanent poison once introduced into the environment".[10]

Similar "keep-the-options-open" rules are used in more mundane ways in energy policy making. Many analysts recommend "open-ended planning" in the sense that "any choice made now must be made in such a way that . . . a later generation, or the same generation at a later date, can reverse the choice and return to the original situation".[11]

This is a principle endorsed even by the United Kingdom's Department of Energy, whose *Energy Policy* review argued that choices "should not prematurely close options" and whose deputy secretary testified at the Windscale Inquiry of the need to "establish a wide range of energy options and maintain a flexible energy strategy which can be reviewed and adjusted if necessary in light of subsequent developments".[12] Such suggestions parallel Barry's analysis of the demands of intergenerational justice "that the overall range of opportunities open to successor generations should not be narrowed".[13]

Keeping options open does not, of course, entail refusing to make any choices at all. Having chosen one path, it will always be costly to shift over to some other; but, while costly, the shift is at least possible. This is all that keeping the options open requires. Thus we may, consistent with that rule, pursue certain options provided our policy is *reversible* — provided we can backtrack if necessary. Such considerations must be superimposed on ordinary decision rules, since expected value calculations overproduce irreversible outcomes even among risk-neutral decision makers.[14]

For an example of the reversibility criterion explicitly at work in policy debates, consider the issue of nuclear power. As regards implementation of the nuclear option, Rochlin recommends "reversibility" as one of two "social criteria" to be used in selecting sites and techniques for disposing of radioactive wastes.[15] It is better, he argues, to place radioactive wastes in deep rock deposits, where we can recover them if something goes wrong and they begin leaking, than it would be to put them where we cannot get them back (letting them melt their way into the polar ice or slide between continental plates in the deep seabed or shooting them off into deep space).

The same sort of reversibility criterion can be used to guide our choice between nuclear and non-nuclear energy options. Elster supposes that such considerations make fossil fuels preferable to nuclear ones.[16] Both entail risks of catastrophic consequences: carbon dioxide discharges from coal- and oil-fired plants might alter the global climate; proliferation of nuclear weapons built with plutonium from nuclear power plants might hasten Armageddon. The difference is that the "hothouse" effect could be reversed. If it really does happen — which is far from certain — then by ceasing to burn fossil fuels we can gradually reverse the damage to the climate. It is a slow process and less certain of success than Elster supposes, perhaps; but at least reversibility looks possible here. Nuclear proliferation, in contrast, seems truly irreversible. Once nations have the technical capacity to build nuclear weapons, they have it forever. Nuclear reactors and their radioactive by-products themselves impose irreversible obligations. As the Oak Ridge team concedes, "One cannot simply abandon a nuclear reactor the way one can abandon a coal-fired plant." It is instead an "unforgiving" technology which, as Kneese worries, "will impose a burden of continuous monitoring and sophisticated management of dangerous material, essentially forever. The penalty of not bearing this burden may be unparalleled disaster. This irreversible burden would be imposed even if nuclear fission were used only for a few decades, a mere instant in the pertinent time scales."[17]

Although the expected-utility rule must obviously be modified to take better account of irreversibilities, it would be inadvisable and, indeed, impossible to make decisions strictly on the basis of the reversibility/keep-the-options-open rule. Usually some options can be kept open only by closing off some others. This, many argue, is the case with energy policy choices. Lovins maintains that "hard" and "soft" energy paths "are mutually exclusive . . . . Commitments to the first may foreclose the second." Other interveners at the Windscale Inquiry claimed that, although reprocessing was justified in terms of keeping the options open, it would actually foreclose them: given the immense capital investment, we will inevitably be stuck with using such plants once we have built them.[18] If some options can be kept open only by closing off others, we must look closely at the likely costs and benefits of each. It would be foolhardy to keep the second, third and fourth options open if the price is foreclosing the first. On the other side, there are surely some options that should be closed forever. One obvious example is the option to initiate a nuclear war. Some, such as Farley, defend the expansion of "peaceful" nuclear programmes on the grounds that "they constitute a hedge against failure of nonproliferation efforts, an assurance that countries which try the nonproliferation option will not be permanently disadvantaged if it fails".[19] But here reversibility looks singularly unattractive. Keeping the option of acquiring and using nuclear weapons open in this way would be clearly indefensible if we ever could find a way to close it irrevocably for everyone. Irreversibility in this case would be a virtue. Similarly, suppose we find a guaranteed way to dispose of radioactive wastes once and for all. Provided we can be absolutely sure of the technique, it would be far better to dispose of these wastes permanently and irreversibly instead of keeping them where we (along with saboteurs, thieves and careless miners) can get at them.[20]

III
Comparing the Alternatives

If forced to make irreversible choices, we should at least do so on the basis of a full survey of all advantages and disadvantages of all available alternatives. Cost-benefit/expected-utility analysis is one way of doing so, but not the only way. This "compare-the-alternatives" principle is enshrined in the economic notion of opportunity costs and was recently brought to the attention of political theorists by Fishkin's discussion of "tyrannical-decision" rules, defined as those which impose gratuitous suffering.[21] This principle of comparing the alternatives is clearly practised in energy debates when various authors routinely compare costs, benefits and risks arising from alternative strategies. And it clearly underlies their criticisms, each of the others, for failing to make *comprehensive* comparisons between *all* the alternatives which are available.[22]

Alongside these very ordinary applications of the compare-the-alternatives principle, we also find a rather more novel one. This assesses the riskiness of a policy in terms of the *increment* of risk it adds to those pre-existing in the status quo, rather than in terms of the absolute value of the risk associated with the policy. Føllesdal's paper on recombinant DNA research, for example, argues that we need not fear a "mad scientist" using these new techniques to unleash a deadly virus on the world, not because that could not happen but rather because there are already enough devastatingly lethal viral and chemical agents available to any given "mad scientist" to do the job. The "added risk" entailed in offering one more way to destroy all life on earth is effectively zero, although of course the absolute risk of that person doing so is frighteningly high.[23]

This principle emerges at various points in the nuclear-energy debate. In Cochran and Rotow's standards for an acceptable system for disposing of nuclear wastes, for example, the radiation hazard to future generations need not be eliminated altogether but only reduced to where it is "comparable to the cumulative risk to all future generations from the original uranium resources from which the radioactive wastes were derived, assuming these uranium resources were unmined".[24] Risk-added reasoning may also underlie our failure to protect future generations adequately from even rather larger risks of leaks of radioactive wastes. The present generation may be willing to forgo nuclear energy in light of such risks to successors "if that action would protect future societies forever. But it would be in no position to control the choices of future societies." And, as long as someone will contaminate their environment anyway, it might as well be us.[25] The risk-added argument is also used by the Oak Ridge team to brush aside concern with the proliferation of nuclear weapons: "We believe that the effect of a moratorium [on civilian uses of nuclear power] adopted only by the United States would be marginal ... because reactors would be available from other countries", and these could supply plutonium for bomb building even if American-built reactors did not.[26]

Wohlstetter aptly replies that "to argue ... that such restrictions would be irrelevant because there are other ways to get a bomb is like opposing inoculation for smallpox because one might also die of bubonic plague. Better to suggest protection against the plague."[27] That the status quo contains other ways for nuclear weapons to spread or for people to be poisoned or irradiated is a criticism of the status quo rather rhan a defence of policies only moderatly increasing those risks. The general flaw of the risk-added approach is that it adopts the status quo as a baseline. Truly comprehensive risk-benefit analysis acknowledges no baseline. We must compare the profiles of *all* the alternatives, the status quo being just one option among many. The fact that an option is no worse than the status quo is irrelevant if there are options available that are significantly better than the status quo.

Finally, we can question the moral relevance of how bad one's alternative opportunities might be. Would we really feel comfortable buying up a bankrupt farmer's land at bargain-basement prices just because we know that there are no other bidders; would we not have a moral duty to pay a "fair market price", even if the market did not force us to do so?

The recent United States Supreme Court decision on the drug Laetrile, thought by some to cure cancer, offers a case in point. Those taking the drug are dying of cancer anyway — their alternatives are grim indeed. Still, we object to other people taking advantage of their sadly restricted opportunity set to peddle drugs which have never been shown safe and effective. The fact that cancer victims have little to lose is beside the point. The court held that the law's protection extends even to the terminally ill.[28] Judging from this case, we do seem to feel that certain things should or should not be done, whatever the competing alternatives look like. The injunction to "compare alternatives" or "inspect opportunity costs "might have distinctly limited applicability if many cases fit into this category.

IV
Protecting the Vulnerable

A further rule, suggested by the last objection to the compare-the-alternatives rule, might require that we give special protection to those who are particularly vulnerable. Someone may be vulnerable to another's actions if he is strongly and directly affected by them; or he may, like the cancer victim or the bankrupt farmer, simply be vulnerable *tout court,* in that he has so restricted an opportunity set as to put him at the mercy of others in general.

Protecting the vulnerable might mean many things. Some environmentalists seem to play on this principle when pointing out that, given modern technology, seals and whales are utterly at our mercy and are enormously vulnerable to our choices. This principle also underlies the practice of judging the acceptability of pollution and radiation hazards by reference to the dose risked by the most exposed group. The most striking application, however, seems to be to future generations, who are peculiarly vulnerable to the effects of our choices, especially in the environmental or energy fields. They would, for example, be strongly and directly affected by our decision to leave them with the radioactive wastes from our nuclear power plants; and, once we have made that decision, they have very little choice but to live with it.

Intergenerational transactions are singularly one-way affairs. Later generations are extraordinarily vulnerable to the choices of earlier ones, but forebears are largely immune to the choices of their successors. This absence of reciprocal relations would, on Hobbesian or Humean accounts, imply that we have no moral obligations toward future generations.[29] But, on other understandings, such asymmetrical power relations are the very stuff of moral obligations: those in a position of dominance have a special obligation to protect those dependent upon them. Such codes clearly underlie the patronal arrangements of peasant societies. Similarly, in our own cultures special vulnerabilities underlie codes of professional ethics and form the basis of loving relationships.[30] Or, again, within the family, the very fact that children are dependent upon their parents gives them rights against the parents. Were we to extend such a principle to policy decisions more generally, we would have to cease discounting the interests of vulnerable future generations. Instead, we would have to count their suffering (from, e.g., our leaky nuclear-waste dumps) at least on a par with our own pains and pleasures.

V
Maximizing the Minimum Payoff

Protecting the vulnerable is recommended on ethical grounds alone. A related principle — maximin, or "maximize the minimum payoff" — is recommended on epistemic ones as well. This rule compares the "worst possible outcomes" of all alternative policies, selecting that policy with the least unbearable consequence should worse come to worst.[31] This decision rule, or something very much like it, is forced upon us whenever any of the three crucial steps in the expected-utility calculus is impossible. That procedure requires that we (a) list all the possible outcomes, and that we then set (b) values and (c) probabilities for each. One or more of these steps is often not feasible. It is often impossible to list all the possible scenarios and outcomes, especially where the "human factor" might be involved (as in the operation or sabotage of a nuclear reactor).[32] Furthermore, it is typically difficult to get reliable probability estimates: objective statistics are unavailable; theories are either too few or too numerous and contradictory; and subjective estimates are unreliable. Finally, it might even be difficult to get a good indicator of the value (or cost) of each of the consequences. Such problems might motivate our reluctance to let a smoker take any risks he wants — we just do not believe he fully appreciates the pains of lung cancer.

Where any of these steps cannot be performed, the expected-utility/cost-benefit/benefit-risk calculations cannot be performed. Instead we must rely upon other types of decision rule capable of functioning without those inputs. Two alternative rules are widely discussed. Føllesdal suggests that "in such cases we estimate the probability of the worst consequence to be 1, and act accordingly", which is the maximin (maximize the minimum payoff) rule. Closely related to this is Savage's "minimax regret rule", requiring us to choose the course of action which minimizes the maximum regret we might suffer.[33] Alternatively, we might follow the suggestion of Arrow and Hurwicz, adopted by the Swedish Energy Commission for assessing nuclear reactors, to choose policies upon the basis of a weighted combination of the best possible and worst possible outcomes which might result from each alternative policy option.[34] If, as Elster argues is the case with alternative energy strategies, all options have roughly the same best possible outcomes, then the Arrow-Hurwicz rule reduces to the maximin rule, and we need only worry about maximizing the minimum possible payoffs.[35] Either of these rules would go far towards avoiding the stringent preconditions for applying the expected-utility rule: while both do require that we know enough about the range of possible outcomes to pick out the worst (or best and worst, or most regrettable), none requires any more information about intermediate possibilities, their costs or benefits; and, most especially, none requires us to know anything whatsoever about probabilities.

Popular resistance to untried technologies with the potential for causing large-scale catastrophes — usually dismissed as a rational risk aversion — might therefore be a wholly rational response to irresolvable uncertainties. It is, I think, clear enough that such uncertainties plague energy choices and render expected-utility calculations impossible. What this may mean for the choice is, however, unclear. The British Department of Energy argued in favour of the nuclear option, saying that "our energy strategy should be robust, producing minimum regret whatever course future events take".[36] Apparently the most regrettable outcome they could imagine is running short of energy. Others draw attention to plausible scenarios with still more regrettable outcomes, ranging from altering the world climate (a possible result of burning oil and coal) to genetic mutations or proliferation of nuclear weapons (a possible result of nuclear power). On balance, it would seem that both nuclear and conventional fossil-fuel power plants are disqualified on this maximin criterion in favour of energy conservation combined with a range of "alternative" strategies relying on solar, wind, wave, geothermal, etc., power.

VI
Maximizing Sustainable Benefits

Another rule directs us to opt for the policy producing the highest level of net benefits which can be *sustained* indefinitely. This contrasts with the directive of ordinary expected-utility maximization to go for the highest total payoff without regard to its distribution interpersonally or intertemporally. Utility maximization looks only to the sum total of benefits and is indifferent to whether they come in a steady stream or all bunched in one period. Considerations of intergenerational equity would demand instead that each generation be guaranteed roughly equal benefits and insist that one generation may justly enjoy certain benefits only if those advantages can be sustained for subsequent generations as well.[37] Following the "maximize-sustainable-benefits" rule would strongly encourage current decision makers to think in *maximin* terms also, since the lowest possible payoff is one that the initial generation must suffer along with everyone else.

The rule of maximizing sustainable benefits has clear and important applications to certain aspects of environmental policy, such as setting permissible levels of fish catches.[38] But it also has significant implications beyond these obvious applications. In the case of energy policy, for example, it decisively favours renewable sources (solar, geothermal, wind, wave, etc.) over utilization of scarce natural resources (such as oil and coal). Uranium falls into the category of scarce resources: "The amount of uranium available to the United States at costs that can be afforded in a LWR [light water reactor] is usually estimated to be 3×10^6 tons. Thus, prima facie, we have enough uranium to support about 25,000 reactor-years of LWRs — say 800 reactors for 30 years."[39] With the "fast breeder", of course, the supply of fissionable plutonium could be rendered virtually inexhaustible, transforming nuclear energy into a sustainable benefit. Then the question is simply which strategy, among those yielding sustainable benefits, yields maximal ones. But there is, more fundamentally, a question of whether the benefits of nuclear energy really are sustainable. The benefits must, remember, be net of costs, which will be increasing throughout time on the most plausible accounts.[40] Even if the energy flow remains constant, the benefits of nuclear power net of these constantly increasing costs will be steadily diminishing.

VII
Avoiding Harm

In all the previous principles, harms and benefits are treated as symmetrical. To avoid a harm is to produce a benefit. We are indifferent between two plans, one generating positive benefits valued at £x and the other avoiding costs of £x which we would otherwise have had to suffer. Avoiding a harm of a certain magnitude is just as desirable (and, indeed, arguably equivalent to) producing a benefit of that same size. The "harm-avoidance" principle denies this symmetry, arguing instead that it is much worse to create costs than it is just to fail to produce equally large benefits.

Initially we might be inclined to run this together with the more familiar "acts/omissions" doctrine.[41] But, in truth, the harm-avoidance principle is not only distinct from that doctrine but is also part of what really underlies its appeal. People who are victims of an immoral *act* – of fraud or criminal assault, for example – are generally worse off than they would have been in the absence of such an act. Where someone merely *omits* to perform a morally desirable act, others are usually no worse off than they were before the omission – they have just lost out on some further benefits they might have enjoyed had the action been performed. Psychological studies show that individuals, when making decisions, generally do weigh losses more heavily than corresponding gains.[42] And ethically it does seem worse actually to harm someone than merely to fail to help him. This sentiment underlies "negative utilitarianism" and the distinction Foot draws between weaker "positive duties" to help people and stronger "negative duties" not to harm them.[43]

Harm avoidance is one of the more important components in environmentalist arguments against reckless interventions into natural processes. Consider, for example, the problem of seeding hurricanes. The hope is that they will thereby lose force before hitting land, and there is every reason to believe that such an action will save many lives and reduce property damage. But there is also a slight chance that seeding the hurricane will make it worse and increase the costs. Some decision-theoretical treatments of the problem, sensitive to the harm-avoidance principle, weight such potential costs more heavily than the costs resulting from just letting the hurricane taking its natural course: the government bears more responsibility for causing such damage than for letting similar damage occur naturally.[44]

Applying this principle to the problem of energy choices would, I think, argue decisively in favour of alternative and renewable sources (solar, geothermal, wind, wave, etc.), combined with strenuous efforts at energy conservation, in preference to nuclear or fossil-fuel generation of power. The worst that can be said against alternative, renewable sources is that they may yield less energy – that they produce fewer benefits. Both nuclear and oil- or coal-fired plants, in contrast, run real risks of causing considerable harm. If we weight the harms much more heavily than the benefits forgone, as the harm-avoidance principle directs, both nuclear and conventional power plants will appear much less advantageous than reliance upon solar, geothermal, wind or wave power combined with energy-conservation programmes.

VIII
Conclusion

The upshot of this discussion is that the standard maximize-expected-utility decision rule has very serious limitations. It is at best a partial response to the range of considerations that should be taken into account by policy makers, especially (but not exclusively) when environmental, natural resource or energy decisions are at issue. At the very least, we would want to modify the maximize-expected-utility rule to take these neglected considerations into account. We might, for example, want to weight outcomes in the expected utility calculus in such a way as:

1. to bias decisions against *irreversible* choices (which may sometimes be permissible, but only after much more careful scrutiny than they receive in the ordinary expected-utility calculus);
2. to bias decisions in favour of offering special protection to those who are especially *vulnerable* to our actions and choices;
3. to bias decisions in favour of *sustainable* rather than one-off benefits; and
4. to bias decisions against *causing harm,* as distinct from merely for-going benefits.

Even after the expected-utility calculus has been modified to meet these further ethical demands, epistemological ones remain. When the logical preconditions for applying that rule are absent, we are logically compelled to fall back on other principles (such as "maximize the minimum payoff") that build on weaker premises altogether. Obviously it would be folly to suppose that all these new rules always converge in their recommendations on particular cases. But as a general rule they would all tend to strengthen the ethical case for environmental protection.

Notes

1. Robert E. Goodin, "Loose Laws", *Philosophica* 23 (1979): 79–96. For another application, see Marshall Cohen, Thomas Nagel and Thomas Scanlon, eds., *War and Moral Responsibility* (Princeton: Princeton University Press, 1974).
2. See Charles E. Lindblom and David K. Cohen, *Usable Knowledge* (New Haven: Yale University Press, 1979); Carol H. Weiss, ed., *Using Social Research for Public Policy Making* (Lexington, Mass.: D. C. Heath, 1977).
3. On cost-benefit analysis generally, see: I. M. D. Little and J. A. Mirrlees

Project Appraisal and Planning for Developing Countries (New York: Basic, 1974); A. K. Dasgupta and D. W. Pearce, *Cost-Benefit Analysis* (London: Macmillan, 1972); and Richard Layard, ed., *Cost-Benefit Analysis* (Harmondsworth: Penguin, 1972). For environmental policy applications see: A. M. Freeman III, R. H. Haveman and A. V. Kneese, *The Economics of Environmental Policy* (New York: Wiley, 1973); Lincoln Allison, *Environmental Planning: A Political and Philosophical Analysis* (London: Allen & Unwin, 1975); and Robert E. Goodin, *The Politics of Rational Man* (London: Wiley, 1976), pt. 4. For critiques, see: Peter Self, *Econocrats and the Policy Process* (London: Macmillan, 1975); Alan Coddington, " 'Cost-Benefit' as the New Utilitarianism", *Political Quarterly* 42 (1971): 320–25; and Laurence Tribe, "Policy Science: Analysis or Ideology?", *Philosophy and Public Affairs* 2 (1972): 66–110.
4. R. M. Hare, "Contrasting Methods of Environmental Planning" and Jonathan Glover, "How Should We Decide What Sort of World Is Best?", in *Ethics and Problems of the 21st Century,* ed. K. E. Goodpaster and K. M. Sayre (Notre Dame, Ind.: University of Notre Dame Press, 1979), pp. 63–78 and 79–92 respectively. Jon Elster, "Sour Grapes: Utilitarianism and the Genesis of Wants", *Beyond Utilitarianism,* ed. A. K. Sen and B. Williams (Cambridge: Cambridge University Press, 1982). Robert E. Goodin, "Retrospective Rationality", *Social Science Information* 18 (1979): 967–90.
5. Robert E. Goodin, "How to Determine Who Should Get What", *Ethics* 85 (1975): 310–21.
6. Richard Zeckhauser, "Procedures for Valuing Lives", *Public Policy* 23 (1975): 454, concludes that "it would be quite rational" for a woman contemplating the threat of breast cancer "to insure no more than the medical expenses" of a mastectomy, since the insurance payment would not restore her breast should it have to be removed.
7. Cf. James Griffin, "Are There Incommensurable Values?", *Philosophy and Public Affairs* 7 (1977): 39–59.
8. Coddington, "New Utilitarianism"; Goodin, *Politics of Rational Man,* 16; Aaron Wildavsky, "The Political Economy of Efficiency", *Public Administration Review* 26 (1966): 292–310. This example further assumes that people will not be compensated for the loss of their homes, or that they cannot be fully compensated by monetary payments for moving from where they grew up.
9. Burton A. Weisbrod, "Deriving an Implicit Set of Governmental Weights for Income Classes", in *Cost-Benefit Analysis,* ed. Layard, pp. 395–428.
10. For further discussion of the reversibility rule, see
(a) M. P. and N. H. Golding, "Why Preserve Landmarks? A Preliminary Inquiry", *Ethics and Problems of the 21st Century,* ed. Goodpaster and Sayre, pp. 175–90,
(b) Clarke's maxim is quoted in Lynton Keith Caldwell, *Environment: A Challenge for Modern Society* (Garden City, N.Y.: Natural History Press, 1970), p. 214,
(c) Carroll L. Wilson et al, *Man's Impact on the Global Environment,* Report of the Study of Critical Environmental Problems, vol. 4, sec. 138 (Cambridge: MIT Press, 1970), pp. 259–63.
11. David W. Pearce, Lynne Edwards and Geoff Beuret, *Decision Making for Energy Futures* (London: Macmillan, 1979), p. 26.
12. Ian Breach, *Windscale Fallout* (Harmondsworth: Penguin, 1978), pp. 27–28.
13. Brian Barry, "Circumstances of Justice and Future Generations", in *Obligations to Future Generations,* ed. Richard Sikora and Brian Barry (Philadelphia: Temple University Press, 1978), p. 243. See also, Barry, "Justice Between

Generations", in *Law, Morality and Society,* ed. P. M. S. Hacker and J. Raz (Oxford: Clarendon Press, 1977), p. 275.

14. Kenneth J. Arrow and Anthony C. Fisher, "Environmental Preservation, Uncertainty and Irreversibility", *Quarterly Journal of Economics* 88 (1974): 312–19; Claude Henry, "Investment Decisions Under Uncertainty: The 'Irreversibility' Effect", *American Economic Review* 64 (1974): 1006–12.

15. Gene I. Rochlin, "Nuclear Waste Disposal: Two Social Criteria", *Science* 195 (1978): 23–31.

16. Jon Elster, "Risk, Uncertainty and Nuclear Power", *Social Science Information* 18 (1979): 371–400.

17. Alvin M. Weinberg *et al., Economic and Environmental Impacts of a U.S. Nuclear Moratorium, 1985–2010,* 2nd ed. (Cambridge: MIT Press, 1979): 79. Allen V. Kneese, "The Faustian Bargain", *Resources* 44 (1973).

18. Amory B. Lovins, *Soft Energy Paths* (Harmondsworth: Penguin, 1977), pp. 26, 59–60; Breach, *Windscale Fallout.*

19. Philip J. Farley, "Nuclear Proliferation", *Setting National Priorities: The Next Ten Years,* ed. H. Owen and C. Schultze (Washington, D.C.: Brookings Institution, 1976), p. 160.

20. Robert E. Goodin and Ilmar Waldner, "Thinking Big, Thinking Small and Not Thinking At All", *Public Policy* 27 (1979): 1–24.

21. James Fishkin, "Tyranny and Democratic Theory", in *Philosophy, Politics and Society,* 5th series, ed. P. Laslett and J. Fishkin (Oxford: Blackwell, 1979), pp. 197–226.

22. Lovins, *Energy Paths;* Weinberg et al., *Economic & Environmental Impacts;* Herbert Inhaber, "Risk with Energy from Conventional and Nonconventional Sources", *Science* 203 (1979): 718–23.

23. Dagfinn Føllesdal, "Some Ethical Aspects of Recombinant DNA Research", *Social Science Information* 18 (1979): 401–19.

24. Thomas B. Cochran and Dimitri Rotow, "Radioactive Waste Management Criteria", mimeographed (Washington, D.C.: Natural Resources Defense Council 1979); Weinberg et al., *Economic and Environmental Impacts,* p. 94.

25. Zeckhauser, "Procedures" p. 439.

26. Weinberg et al., *Economic & Environmental Impacts,* p. 56.

27. Quoted by Czech Conroy, *What Choice Windscale?* (London: Friends of the Earth, 1978), p. 57. See further, Albert Wohlstetter, "Spreading the Bomb Without Quite Breaking the Rules", *Foreign Policy* 25 (1976–77): 145–79.

28. *United States* v. *Rutherford* 442 U.S. 544, 551, 555, 558 (1979) came to the U.S. Supreme Court on appeal from the Tenth Circuit Court of Appeals, which "held that the safety and effectiveness terms used in the statute have no reasonable application to terminally ill cancer patients". Since those patients, by definition, would "die of cancer regardless of what may be done", the court concluded that there were no realistic standards against which to measure the safety and effectiveness of a drug for that class of individuals. The Court of Appeals therefore approved the District Court's injunction permitting use of Laetrile by cancer patients certified as terminally ill. The U.S. Supreme Court unanimously overturned this ruling. Mr Justice Marshall, for the Court, writes, "Only when a literal construction of a statute yields results so manifestly unreasonable that they could not fairly be attributed to congressional design will an exception to statutory language be judicially implied. Here, however, we have no license to depart from the plain language of the Act, for Congress could reasonably have intended to shield terminal patients from ineffectual or unsafe drugs .... Since the turn of the century, resourceful entrepreneurs have advertised a wide variety of purportedly simple and pain-

less cures for cancer, including lineaments of turpentine, mustard, oil, eggs, and ammonia; peatmoss; arrangements of colored floodlamps; pastes made from glycerin and limberger cheese; mineral tablets; and "Fountain of Youth" mixtures of spices, oil, and suet . . . . This historical experience does suggest why Congress could reasonably have determined to protect the terminally ill, no less than other patients, from the vast range of self-styled panaceas that inventive minds can devise."

29. Barry, "Justice Between Generations", idem, "Circumstances of Justice".
30. James C. Scott, *The Moral Economy of the Peasant* (New Haven: Yale University Press, 1976); Bernard Williams, "Politics and Moral Character", in *Public and Private Morality*, ed. S. Hampshire (Cambridge: Cambridge University Press, 1978), p. 56; John R. S. Wilson, "In One Another's Power", *Ethics* 88 (1978): 299–315.
31. Maximin implies avoiding the worst state of the world all round, whereas protecting the vulnerable implies avoiding the worst consequences possible for certain target groups. It is, of course, possible to build a similar distributive focus into maximin (as does Rawls, for example): but it is important to realize that we *are* building something more into the rule when we do that.
32. Robert E. Goodin, "Uncertainty as an Excuse for Cheating Our Children", *Policy Sciences* 10 (1978): 25–43.
33. Føllesdal, "Ethical Aspects", p. 406; L. J. Savage, *The Foundations of Statistics* (New York: Wiley, 1954).
34. Kenneth J. Arrow and Leonid Hurwicz, "An Optimality Criterion for Decision-Making under Ignorance", in *Uncertainty and Expectations in Economics*, ed. C. F. Carter and J. L. Ford (Oxford: Blackwell, 1972), pp. 1–11; Swedish Energy Commission, *Mijöeffekter och risker vid utnyttjande av energie* (Stockholm: Liber Förlag, 1978).
35. Elster, "Risk, Uncertainty and Nuclear Power".
36. Breach, *Windscale Fallout*, p. 28.
37. Talbot Page, *Conservation and Economic Efficiency* (Baltimore: Johns Hopkins University Press for Resources for the Future, 1977). Even the goal of maximizing (undiscounted) utility usually – although not always – prohibits destruction of "interest-bearing resources . . . like crop species, fish species, draft animal species, topsoil, genetic variation, etc . . . . which are such that their capacity to supply energy for future consumption is not decreased by the utilization of some of the energy they supply . . . . Interest-bearing resources renew themselves and provide a bonus for us", and total utility derived from them is usually maximized by skimming off the interest and leaving the capital intact. This, once again, amounts to extracting the maximum sustainable yield from the resources, as argued in Mary B. Williams, "Discounting versus Maximum Sustainable Yield", in *Obligations to Future Generations*, ed. Sikora and Barry, p. 170.
38. Arild Underdal, *The Politics of International Fisheries Management* (Oslo: Universitetsforlaget, 1980).
39. Weinberg et al., *Economic and Environmental Impacts*, p. 82.
40. For a survey of these arguments, see Robert E. Goodin, "No Moral Nukes", *Ethics* 90 (1980): 417–49.
41. G. E. M. Anscombe, "War and Murder", in *War & Morality*, ed. R. A. Wasserstrom (Belmont, Calif.: Wadsworth, 1970), pp. 42–53, Jonathan Glover, *Causing Deaths and Saving Lives* (Harmondsworth: Penguin, 1977), pp. 86–112.
42. Frederick Mosteller and Philip Nogee, "An Experimental Measurement of Utility", *Journal of Political Economy* 59 (1951): 371–404, Anatol Rapaport

and T. S. Wallsten, "Individual Decision Behavior", *Annual Review of Psychology* 23 (1972): 131–76.

13. Philippa Foot, *Virtues and Vices* (Oxford: Blackwell, 1978), pp. 28–30. Much the same distinction is drawn by Charles Fried, *Right and Wrong* (Cambridge: Harvard University Press, 1978), Ch. 2, although he justifies it in terms of "intentions". On negative utilitarianism see, H. B. Acton, "Negative Utilitarianism", *Proceedings of the Aristotelian Society (Supplement)* 37 (1963): 83–94; and Barrington Moore, Jr., *Reflections on the Causes of Human Misery* (Boston: Beacon Press, 1970).

14. R. A. Howard, J. E. Matheson and D. W. North, "The Decision to Seed Hurricanes", *Science* 176 (1972): 1191–1202.

Political Representation
for Future Generations
GREGORY S. KAVKA and VIRGINIA WARREN

Theories of democracy disagree over whether political representatives in a democratic system represent (or should represent) interest groups, their district's citizens, political parties, or the nation as a whole. Yet these diverse theories share a common presupposition: that whichever groups are the locus of representation, it is their *presently existing* members (or the interests of their presently existing members) that are represented.[1]

It is neither surprising nor objectionable that *descriptive* theories of representation should make this assumption. For, in current democratic systems, no special institutional mechanisms exist to secure representation of future people's interests, and representatives naturally focus their attention on promoting the interests of those who have the power to vote them into, or out of, office; that is, present citizens. Yet the same assumption is made, without supporting arguments being offered, in virtually all *normative* theories — theories concerning those whom democratic representatives *ought to* represent. Here also the soundness of the assumption may be questioned. For it rules out the possibility of the interests of the nation's future citizens — whose lives will be critically affected, for better or worse, by present government action — being directly represented in the democratic political process.

The idea that these future people should be so represented is alien to our traditional way of thinking, so much so that it will strike many as being absurd. Below, we shall attempt to dispel this aura of absurdity, by carefully dealing with a number of significant objections that might be raised against the notion of political representation for future generations. Before beginning this task, however, it would be useful to explain why the gap we claim to have found in democratic theory has not been sufficiently noted before, and why it now poses a serious problem that demands attention.

Scientific knowledge and technology have grown spectacularly in this century. The development of modern industry and technology has given recent generations the capacity, far surpassing that of earlier generations, to do severe harm to the life prospects of future people, by seriously degrading the natural environment, using up vital natural resources, or destroying civilization by nuclear war.[2] Further, applications of modern scientific knowledge and technology in agriculture and medicine have accelerated population growth worldwide. The resulting overpopulation now threatens to destroy elements of the natural and social environment which must be preserved if future people are to live decent lives.[3] At the same time, advances in the natural and social sciences (including the use of computers) have made us more aware than ever before of how our present activities are likely to affect the life prospects of future generations, and of how we might alter these activities to improve the likely situation of future people. Thus, the same tool — modern scientific and technological knowledge — has given us both unprecedented power to injure posterity, and unprecedented awareness of the nature of that power and the possible means of controlling it or its undesirable effects. In short, because the choices which democratic societies make now have enormous impact on future generations, and because we collectively know this and can make better than random guesses about the long-range outcomes of different choices, it is appropriate at this time to raise the question of political representation for future people.

I
Can Future Generations Be Represented?

We shall assume that it makes sense to speak of those currently living having obligations to future generations, and that we do in fact have certain obligations of this kind, for example, to preserve a planetary environment that can support human life. Objections to this, based on the claim that we cannot have obligations to those who do not exist, or to those whose lives will not overlap our own, are sometimes raised. But these objections have been adequately rebutted elsewhere.[4]

We wish here to consider objections to the idea that a democratic nation's obligations to future generations can be, or should be, fulfilled (in part) by according future generations representation within the political process.[5] Setting aside more practical objections which are treated in section III below, the main theoretical objections to political representation for future generations fall roughly into two categories. Firstly, there are objections based on the existence of other classes of persons, besides future people, whose interests are gravely affected by decisions made in the political process, but who are not represented in that process. These objections shall be discussed in section II. Secondly, there are a number of objections of the form, "It does not make sense to talk of representation for future people. The nature of representation is such that future generations *cannot* truly be represented." In this section, we shall deal with objections of this sort, and shall attempt to show that none is convincing.

The general issue concerns the *prerequisites* (i.e., necessary conditions) of one party representing another. In particular, two questions arise. What features must the members of a class of entities possess, in order for it to be logically possible for someone to represent them? And what must the relationship be between one party and another, if the latter is to be a genuine representative of the former? Different answers to these questions yield different versions of the claim that the notion of representation for future generations is incoherent.

Firstly, it might be suggested that present existence is itself a prerequisite of being represented. If this is so, it necessarily follows that future generations cannot be represented. When put in the rhetorical form "How could you possibly represent what does not exist?", this suggestion may sound plausible. But it is clearly refuted by some of the legal practices of modern nations, in which lawyers, trustees of estates, etc., often represent deceased persons, or the interests of particular future persons (e.g., unborn descendants for whom trusts are established).

Though readily disposed of, the above objection is at the root of two others. Each involves the claim that there are certain prerequisites of being represented that future people fail to satisfy because they do not exist yet. If one thinks of paradigm cases of representation – such as a lawyer representing his client or a congressman representing his district's voters – it may seem that choosing one's representative, and instructing or advising him, are prerequisites of being represented. But again, if we take into account the range of representation within modern legal systems, we see that this is not so. Children, and mentally incompetent adults, are often legally represented by parents or guardians they have never chosen; while indigent criminal defendants are frequently represented by public defenders appointed by the court.

As to instructions and advice, young children and certain incompetent adults are unable to supply these to their legal representatives. And deceased persons cannot advise or instruct their legal representatives with regard to contingencies not forseen by them while they were alive.

The positive doctrine implied by these legal examples is one suggested by Joel Feinberg:[6] that the main, and perhaps sole, prerequisite of being representable is *having interests,* being capable of faring well or faring ill. (As the representability of deceased persons indicates, we must interpret Feinberg's notion of "having interests" in the atemporal sense of having interests at some time — past, present, or future.) Acceptance of this plausible "interest principle", as Feinberg calls it, does not, however, immediately disarm all conceptual objections to political representation for future generations. One counter-move is to claim that representatives not selected and/or instructed by a group may, at best, represent that group's interests, but not the group itself; or, at least, they cannot be the group's political representatives for, in order to be a political representative of a group, one must be selected for that role by the group. In our view, though, this counter-move is merely a nonsubstantial verbal manoeuver. We are interested in defending the substantive proposal that, within democratic governments, there be specially designated representatives charged with protecting and promoting the interests of future generations. Whether they are called "political" representatives, and whether they are regarded as representing future people or, instead, their interests (if this distinction even makes sense) is of no consequence.

A more serious objection to the interest principle concerns the tasks which a representative must carry out. It may be expressed as follows. If, as the interest principle implies, the essence of representing a group is to protect, promote or speak for its interests, then future generations cannot be represented. For how is a "political representative" to know future people's interests, if future people are not around to tell him what they want, and he does not know what they, their lives, and their living conditions will be like? Further, even if he did know what future people's interests will be, he would not know how to act at present to promote these interests, given how uncertain the long-term effects of government policies and programmes are.

It may be true that if one party has no reasonable beliefs at all about what another party's interests are, or has no idea at all about how to promote what he believes to be the other party's interests, it would be impossible for the former to represent the latter, except by acting on the latter's instructions. But surely we do know a singificant amount about the interests of future people, for at least a large number of generations. These people will have the same (or very similar) biological needs for food, clean air and water, shelter, etc. that we have, and it will be in their interests if these things, and the means of procuring them — including arable land, reliable supplies of energy, and various minerals — are available.[7] And while there are grave uncertainties involved in predicting the long term effects of present social actions and policies, we can make reasonable and nonrandom projections. We can predict, for example, that while future technological developments *might* alleviate our main environmental problems, it is likely that pursuing policies which degrade our biological environment — for example, those which emphasize nuclear fission as a main power source, or which allow industrial pollution of our air and water — will have a marked negative impact on the quality of life for some (or all) future generations. The ability to make such probabilistic, nonrandom judgments about the effects of policies on the interests of a given individual or group is all that is necessary for representation. In particular, certainty, or something approaching it, is not required. If it were, paradigm cases of representation, such as elected congressmen representing their constituents (or lawyers representing their clients, or even people representing themselves), would not qualify as genuine representation! For typically, congressmen do not know with certainty, or near certainty, what the interests of their constituents are, or how various proposed policies would affect the fulfillment or frustration of those interests.[8]

To summarize, only two plausible logical prerequisites of representation have emerged from our brief discussion. Firstly, the represented party must (atemporally) have interests. Secondly, the representative must either be instructed by the represented party, *or* must know enough about the likely interests of the represented party and the means of promoting those interests, to be able to make better than random judgments about how alternative policies are likely to affect the interests of the represented party. But surely both of these prerequisites can be satisfied in the case of present people representing future generations within a democratic political system. For future people will have interests,[9] and we can have well-grounded and reasonable beliefs about some things that can now be done to protect those interests. If this is so, there are no sound reasons to suppose that future generations cannot, in principle, be represented in a democratic political system.

Why Future Generations Should Be Represented When Other Affected Groups Are Not

We have argued that the idea of political representation for future generations is, indeed, coherent. Before considering how, in practical terms, future generations might be represented, we wish to examine a major theoretical objection to the idea of such representation. It is this: why should the interests of future generations be represented in a democratic nation's political process, when those of certain other groups are not? For (the objection continues), if the moral reason for including the interests of future generations is that current government actions will seriously affect their welfare, then, to be consistent, the interests of *all* those seriously affected by such decisions must be represented. Hence, the challenge is to identify morally relevant differences between other affected, but unrepresented, groups and future generations. Four groups which are (or may be thought to be) both unrepresented and seriously affected by present government decisions are: children, the future selves of current citizens, past generations, and the people of other nations.

Children

Let us first consider children (i.e., minors). In democratic theory, the standard way of ensuring that a group's interests are looked after is to allow members of that group to vote. But, in order to effectively further one's own interests through voting, one must possess a certain degree of knowledge and judgment; minimal competence in this respect is usually measured by age, as a simple approximation, although other means could be used. Since children, especially young children, generally lack the competence to vote, their interests ought to be represented in other ways. The question here is which of two sorts of representation children should have. Is representation by persons or groups whose primary responsibility is to promote the interests of children specifically — what we shall call *direct representation* — required? Or would it suffice to have children's interests represented *indirectly,* by persons representing some other group (e.g., adult citizens) whose members are presumed to care about children's interests?

We contend that direct representation of children's interests is not necessary. For, parental love and concern is one of the strongest forces in human society. As things stand, parents (and other adult relatives and friends) — through their own votes, participation in interest groups, etc. — indirectly represent their children's needs. However, we concede that there are two imperfections in the current practice of letting parents indirectly represent their children's interests.

The first flaw is that some children's interests will receive more representation than others, and this may be seen as unjust. For, children in large families have (at most) two parents, just as children in small families do. So, "one man, one vote" — in this case, one proxy vote — is violated.[10] This would not, in itself, be so unfair, if all groups were affected equally. Unfortunately, the poor (who are disproportionately members of minority groups) tend to have more children per family, and more one-parent households. Moreover, poor adults are less likely either to vote or to contribute money and time to political interest groups. So, children do not now receive equal (indirect) political representation, through no fault of their own.

A second problem is that indirect representation of children's interests is not as effective as one might wish. Consider some examples from the United States: government-supported daycare programmes are not readily available (despite the dramatic rise in two-worker families and "latch-key" children); doctors, etc., have, only recently, been legally required to report cases of child abuse, voters are increasingly turning down school-board requests for desperately needed funds; and, recently, the unemployment rate of teenagers has increased disproportionately, with little being done to remedy the situation. The interests of children, it would seem, should receive more attention in the political process than they do now.

Still, existing children have this one huge advantage over future generations, especially over distant generations: we care *much* more about the former. We see Susie and Johnny; we touch them. A politician need only say the magic words, "for the sake of our children", for a concrete and vivid picture of our own offspring (or of those of people we care about) to appear before us. Our imaginations usually fail to conjure up details about future persons, and so we generally fail to identify with them and their interests.[11]

In short, children are now indirectly represented in the political process, although the representation may be unequal and less than fully adequate. But, because we care about existing children so much more than we do about future generations, there is little chance that the needs of the former will be grossly neglected; the needs of the latter all too frequently are. This difference justifies our view that direct representation is needed for future generations, but not for children.

Later Selves

Our own *later selves* are a second group of entities that are substantially affected by current government policies, but are not directly represented in the political process that determines those policies. This sounds odd, but it is merely an abstract and picturesque way of expressing a commonplace observation. Namely, that in later periods of our lives, we will be affected by government policies and programmes adopted now, but the people we will have become at that time will not have control over the (present) adoption and implementation of those programmes. Suppose, for example, that with no thought of the morrow beyond the next election, current representatives enact a social-security system which goes bankrupt, leaving us penniless upon our retirement thirty years hence. Then our later selves (i.e., ourselves thirty years from now) will have been harmed by their lack of present representation.

Of course, our later selves are indirectly represented by our present selves. For individuals can, and do, consider their long-term as well as short-term interests in deciding what candidates, programmes, and parties to support. It is, however, a notorious fact about human nature, that people tend to *overdiscount* long-term gains and losses relative to short-term gains and losses. (That is, they tend to assign a lesser relative value to future returns than can be justified solely on the ground of the greater uncertainty of future returns.[12] Further, politicians in democratic states, who are elected for relatively short periods and who are judged by voters largely in terms of the immediate results of their actions, also have strong incentives to overdiscount the future in the policy-making process. It is therefore inevitable that the interests of our own later selves will be, in effect, under-represented in the democratic political process. Yet, no one is so foolish as to propose direct representation for our future selves. Indirect, though imperfect, representation through our present selves seems to be all that is required. Why then should we treat future people better than our own future selves, and accord them direct, rather than indirect, political representation?

There are three very good reasons for doing so. The first is parallel to our main reason for saying that direct representation is needed for future generations more than it is for children. It seems certain that indirect representation of future generations will be considerably *less effective* than indirect representation of our later selves, because of the great strength of our concern for our later selves. Nearly all of us identify more strongly with, and care more about, our own later selves, than future generations. As a result, in our political behaviour, we will be more attuned to protecting our own long-term interests than the interests of future people.

Secondly, the standard discounting techniques employed in public-policy analysis follow the economic procedure of compounding. Thus, for example, at a discount rate of fifty per cent per year, $1 now is valued the same as $2 next year and as $4 two years hence. One result of this procedure is that if the discount rate is too high, the gap between the actual value of future returns, and their value discounted at an excessive rate, will increase over time. So, later harms and benefits will be undervalued as compared to earlier ones. But effects of present policies on future generations will generally occur later than effects of those policies on our own later selves. Thus, if voters and government officials rely, in the political process, on policy analyses that use too high a discount rate, the interests of future generations will be under-valued much more than are our own future interests. And, as Mary Williams has persuasively argued in the case of renewable resources, the standard analyses that influence policy do tend to employ unjustifiably high discount rates.[13]

Even if these two reasons did not apply, however, and future generations were no more under-represented than are our later selves, there would still be a strong reason for providing direct representation for the former, but not the latter. For, if the interests of our future selves (i.e., our own future interests) are harmed by under-representation due to overdiscounting, this is our own (collective) fault. It is we who make the mistake of downgrading the importance of the long-term conse-quences of our present policies, and it is we who later pay the penalty. But if future generations suffer because we failed to give sufficient weight to their interests, they are in no way responsible for this. This latter situation is unjust, and calls for a remedy, in a way the former does not.

In case this does not seem obvious, consider the following example. Some cigarette smokers continue to smoke because they overdiscount the possible deadly, but temporally distant, effects of smoking (e.g., contracting lung cancer). Since it is the individual's own welfare that is at risk, the state has no right to prohibit his smoking. However, since there is evidence that smoke inhaled from others' cigarettes can cause disease, if the same discounting attitude inclines a smoker to smoke in close proximity to others, the state may justifiably forbid his doing so.

The same principle (a version of Mill's liberty principle) governs both these cases and those concerning political representation for later selves and future generations: conduct, based on overdiscounting, that is harmful to ourselves should generally be permitted; but such conduct that is harmful to others should be prevented. This principle, together with the two reasons cited above, provides ample grounds for concluding that advocacy of direct political representation for future generations need not commit one to favouring direct political representation for our later selves.

Past Citizens

Consider now a third group: past (i.e., dead) citizens of a nation. They certainly lack direct political representation, but can the nation's present citizens have any influence on their interests now? Some philosophers would say yes, adopting Aristotle's view that one's interests or happiness can be affected by events after one's death.[14] (As, for example, when one's life work is carried to fruition, or abandoned, by one's survivors, or when one achieves posthumous fame or infamy.) If this is so, the interests of past citizens who very much wanted their country to do or to achieve certain things in the future, could be substantially affected by current policies.

Nonetheless, the considerations favouring representation for past and future citizens are not on a par. For, even if we accepted the controversial doctrine that dead citizens are, in some sense, "harmed" by future events not turning out as they had desired, this is harm of a much less serious sort than that suffered by a person who experiences the frustration of his main desires.[15] And future generations may well suffer, *and experience,* harm if they are not adequately represented in present political decision making. It might be noted, in addition, that if now-dead citizens once had desires for the future well-being of their country, it is likely those desires concerned the indefinite future, rather than the time period extending only to the present. Hence, insofar as direct political representation for future generations would promote the well-being of the nation's future citizens, it would likely satisfy dead citizens' desires as well.

We may expand briefly on this last point, by suggesting that we may even be obligated *to past generations* to promote the well-being of future generations, and to adopt the necessary means of doing so (including, possibly, according direct political representation to future generations). To see how this could be so, it will be useful to describe and to give a label to an important class of obligations that has been largely ignored by philosophers. Suppose that person A does person B a special service so that, as a result, B owes A a debt of gratitude. An especially appropriate way to pay off that debt, would be for B to perform a similar service for A. But suppose, as is often the case, that A does not need the service or does not want it. Often what A wants instead is that B provide a similar service to others who need it, just as he has provided it to B. Then we may say that B has a *(prima facie) redirected obligation of gratitude* to provide the service to an appropriate beneficiary, or group or beneficiaries (e.g., those specified by A, and/or those who especially need the service).

We are familiar with such obligations in ordinary life. A mother may make extraordinary sacrifices on her child's behalf and, when asked by the child how this can be repaid, may reply either (i) that it needn't be repaid because the parent is only paying back her parents for bestowing similar benefits on her, or (ii) that the child should reciprocate by doing the same for his or her children. Underlying both replies is the recognition of a redirected obligation of gratitude. Similarly, someone who has benefited from past medical research may volunteer to participate in a dangerous or painful experiment as a way of repaying the debt. He does this even though he realizes that the people helped are not those whose sacrifices helped him.

A possible new source of obligations to posterity emerges, when the existence of redirected obligations of gratitude is recognized. For, some of our ancestors undertook various economic, military, and psychic burdens, partly for our benefit, and they likely did so with the hope and expectation that each succeeding generation would shoulder similar burdens to help those coming after. Thus, we may have redirected obligations of gratitude to aid future generations. If so, consideration of our responsibilities to past citizens supports, rather than calls into question, the appropriateness of political representation for future generations.

Foreigners

Jokingly, a Frenchman who was interviewed on radio prior to the 1980 United States presidential election said that the French ought to vote in this election, since they will be affected so greatly by the outcome. His premise was correct. The decisions of many democratic nations concerning war, food production, use of natural resources, storage of toxic wastes, foreign aid, etc., do seriously affect the peoples of the world. If we are right that future generations' interests should be directly represented and, if no relevant difference can be found between foreigners and future generations, then we would be committed to the direct representation of the interests of foreigners (perhaps even by the ballot, since they are competent to vote) in the political systems of these nations.

We have argued above that the interests of other affected groups — children, our later selves, and past citizens — are well enough represented in a democratic system through indirect means. Is this also true of the interests of foreigners? They do receive indirect representation, to a degree. In the United States, for example, citizens (ethnic groups, in particular) sometimes indirectly represent the interests of foreign peoples, although not in proportion either to the number of foreigners affected by United States decisions or to their need. In addition, foreign cartels (e.g., OPEC), other governments, and international interest groups (e.g., Oxfam) are often able to persuade, or to force, a government to recognize some of the interests of foreigners. Yet, when the overall picture is reviewed, we concede that the interests of affected foreigners are not fully represented through indirect means. Still, there are several significant differences between foreigners and future generations.

Firstly, the interests of foreigners are, in general, currently *much better* represented through indirect means than are those of future generations. Why is this so? We do indeed find it difficult to identify with starving children in far-off countries. But charities succeed in making the lives of geographically remote persons somewhat real to us, by placing full-page advertisements bearing the sad face of little Marika, alongside Marika's letter of thanks to her sponsor. In contrast, we cannot photograph the future (except metaphorically, through art), nor do future people write letters. Hence, it is qualitatively more difficult to identify with future generations, and to take their interests to heart.

Before going further, we should consider what would happen were a democracy to give direct representation to affected foreigners. Such a step would, in fact, be tantamount to destroying that nation's sovereignty.[16] For government actions concerning a particular country would be determined by the residents of that country *and* by affected residents of Brasil, India and Upper Volta. Thus, the government of each democracy would be a United Nations, of sorts. Moreover, local control at every level — state, county, city — could become virtually extinct. And, if *every* nation adopted the principle of direct representation for all affected individuals, then nations would cease to be the basic unit of worldwide political organization; nations would be obsolete, as feudal fiefdoms are now. Our second reason for differentiating between representing the interests of foreigners and future generations is, then, as follows. To put the former into practice would require the abandonment of the system of sovereign states and, hence, is *totally* unrealistic.[17]

Our third reason is that even if direct representation of foreigners were practical, it might not, on balance, be morally desirable. For within the current system of sovereign nations governments can often, through international diplomacy, influence those actions of other nations which seriously affect their citizens' interests. And this system may have an adequate moral basis in the right of self-determination (automony) or in general considerations of utility. If so, we would not be obligated to replace the current system with an alternative one in which the interests of all affected individuals are represented in the making of "national" political decisions.

In conclusion, we have shown that there are significant differences between future generations and other affected but unrepresented groups; and so our proposal to grant direct representation to the interests of future generations, but not to those of other groups, is not arbitrary. In the next section, we hope to show how the interests of future generations *could* be directly represented within sovereign democratic nations.

III
Is Representation for Future Generations Practical?

Having rebutted the main theoretical objections to direct political representation for future generations, we must face the objection that our idea is impractical, too impractical to be taken seriously. In response, we wish to distinguish two senses in which a change in political representation can be impractical: it can be *unacceptable* to the public and to government officials, or it can be *unworkable* within the framework of an already established set of political institutions. We concede that our proposal is, at present, quite impractical in the first sense. It is likely to seem unacceptable to most people because it is novel, sounds radical, has never been tried anywhere, and because it is not clear how or whether it would work if put into practice. In the hope of reducing such skepticism, we shall now broadly sketch two concrete alternatives which do not seem so radical as to be unworkable within the framework of existing democratic institutions. For simplicity, we shall confine our attention to the democratic system with which we are most familiar, that of the United States. The analogues of our proposals for parliamentary systems should be clear.

Despite the curious ring to the title "Representative of Future Generations", when considering political representation it is natural to think first of the legislature. We argued, in section I, that it is theoretically possible to represent the interests of future people, since being represented requires neither that one currently exist, nor that one choose and instruct one's representative.

The more troublesome issue is to determine how many such representatives there should be. While proportional representation is the rule for existing citizens, there are two reasons that, in the case of future citizens, this rule must be broken. Firstly, no one knows how many future citizens will actually exist, or even for how many generations the nation will continue. (In fact, the number of future people might differ according to whether or not we did directly represent their interests.) Secondly, if we assign representatives proportioned to an estimate of the number of future people there are likely to be, representatives of future generations would vastly outnumber those of present citizens. We think it doubtful that present people are morally obligated to submerge their interests to this extent; and, in any case, they certainly would not do so. So, on grounds both of practicality and of morality, a decidedly less-than-proportional representation for future generations would be required.

A modest and workable form of less-than-proportional representation would be to have a handful of representatives for future people, or even a single one in each house of congress. Their function would be to further the interests of future people by participating in debate, working out compromises, proposing legislation, serving on committees, voting, etc. Their impact could be appreciable, despite their small numbers.

There are several possible methods of choosing such representatives. They could be appointed by the president, the only elected official (besides the vice president) who represents the country as a whole. Or, a national election could be held, in which existing citizens chose from among candidates nominated by various political parties.

It might be objected that however such representatives were selected, they might be opposed to giving any (or much) weight to the interests of future generations. Their opposition might either be kept secret during the campaign, or it might even be the cornerstone of their campaign! We admit that these things could happen. But, we doubt that this objection is critical, since our whole political system is alive with similar possibilities. The president could select Supreme Court Justices who did not believe that judges should follow the law, or a C.I.A. director who wished to disband that agency. In fact, citizens could elect to office a slate of anarchists who were bent on dismantling the government. The most that this objection shows is that unless a sufficient amount of public approval exists, arrangements for legislative representation of future generations will not accomplish their purpose.

A second alternative is to have the interests of future people represented within the executive branch, either by a cabinet member, or by a director (comparable to the F.B.I. director) whose tenure was not necessarily tied to a specific administration. This alternative might prove to be more expensive than having legislative representatives, since a whole department or agency would be created. But, then, a department or agency could also conduct research, help to educate the public, have consultants to supply information to any government officials needing it, and issue reports on the predicted effects of government action on the welfare of future generations (analogous to current environmental impact reports). Having a separate body would be preferable, we think, to having "future generations officers" (on analogy to affirmative action officers) spread throughout various agencies. For it is psychologically difficult to be a single outsider, a devil's advocate, in a bureaucratic organization; one is likely to lose sight of one's separate goal, or to be ignored and ineffective.

It surely will be objected to either of these alternatives that adding more government is simply too expensive. We reply that our proposal makes it someone's primary responsibility to call attention to how decisions would affect future generations, and that such a voice need not be expensive to be heard. Naturally, in adding either a few congressmen or a separate department or agency, care must be taken not to waste money. Still, implementing either alternative would cost something. But, there are economic, as well as moral, reasons for doing so. In the long run, it would probably cost more money to ignore future persons' needs than to take care of them (e.g., cleaning up pollution is usually more expensive than preventing it). There are also other long-term costs of ignoring future generations which are not, or not merely, financial. War can result from short-sighted policies about food production, population, energy sources, etc.; and war is paid for not only in money, but also in death and suffering and, less tangibly, in the stifling of progress in the sciences and arts.

One final objection to our thesis is that representatives of future generations are not needed, because the welfare of future people is already being taken care of by some other group — by private enterprise, by existing government officials and agencies, and by special-interest groups. Firstly, we think it not at all realistic to expect action from private enterprise. For, businesses operate for profit, and it is generally not profitable for a business to care about events occurring after it no longer exists.[18] The situation is even more dire, however, since businesses frequently do not act with much foresight even to protect their own interests in the not-so-distant future. For example, American automobile companies for many years fought strenuously against government regulations designed to increase gasoline mileage. Secondly, elected officials tend also to be relatively short-sighted, in order to gain re-election. Thirdly, existing government agencies, such as the Nuclear Regulatory Agency and the Environmental Protection Agency, do handle some of the work we are claiming needs to be done. But much more ought to be done. Also, a central department could develop a comprehensive policy, so that the efforts of existing agencies are coordinated and do not undercut each other. In addition, we think that the debate over the environment, etc., is at present often distorted. We frequently hear that we must choose between the interests of a few intrepid backpackers who would use the wilderness, and the interests of the many who would not. That is, the interests of future generations are often downplayed. If, however, there were official representatives to place those interests in the forefront of the debate, the attention of politicians and the general public could be captured. Fourthly, interest groups, such as the Sierra Club and Zero Population Growth, are not now, and not soon likely to be, capable of adequately defending the interests of future people, nor are these interests their sole concern.[19]

Having outlined two workable alternatives — adding a few congressmen, or a special department or agency, to represent future generations — we would like to comment on the public acceptability of these alternatives. The following is a false dilemma: that direct representation for future generations is either impossible (because people are not concerned about the well-being of future generations) or else superfluous (because people are concerned, and so they already indirectly represent the interests of future generations adequately). A middle ground may eventually emerge, wherein a sufficient number of people are sufficiently concerned to effectively institute direct representation, although many others still do not identify with, or think about, future generations.

To raise public awareness about the effects of our political decisions on future generations, to the level required for changing the political system as we have suggested, would take the combined efforts of concerned parents, educators, scientists, artists, etc., over a period of time. While the task is formidable, we should remember that, a mere decade ago, the public was hardly aware of environmental issues (concern about which largely overlaps with concern for future generations). We do think that direct representation of the interests of future people is needed, and would be workable in either of the forms we have proposed. But, we would be pleased indeed if, a generation from now, people cared so much about future generations, that indirect representation would suffice.

Notes

1. See Neil Reimer, ed., *The Representative: Trustee? Delegate? Partisan? Politico?* (Boston: D. C. Heath, 1967); and Hanna Pitkin, *The Concept of Representation* (Berkeley and Los Angeles: University of California Press, 1967), in which the major theories of representation are discussed, and in which no mention is made of representation of future generations (or their interests).

2. In earlier times, people possessed (though to a lesser degree than now) the capacity to seriously harm future generations *deliberately* (e.g., by burning forests, destroying crop land). In modern times, however, future generations may suffer, as side-effects of our productive activities, substantial harms we do not want to impose on them. Thus, the possibility of our harming future people now depends less on the unlikely motive of collective malevolence toward them, and more on the likely motives of carelessness and selfishness. Hence the present danger, if ignored, is far greater.

3. See, for example, Donella H. Meadows et al., *The Limits to Growth* (New York: Universe, 1972). For a contrasting view, see Herman Kahn et al., *The Next 200 Years* (New York: William Morrow, 1976).

4. For example see, Richard and Val Routley, "Nuclear Energy and Obligations to the Future", *Inquiry* 21 (1978): secs. 2–3. Cf. Gregory S. Kavka, "The Futurity Problem", in *Obligations to Future Generations*, ed. Richard Sikora and Brian Barry (Philadelphia: Temple University Press, 1978).

5. For the purposes of discussing representation within a national political system, we shift focus from our *generation's* obligations to posterity, to a

nation's obligations to its posterity (i.e., the obligations of a nation's present citizens to its future citizens). Cf. p. 000, n. 16 in this paper.

6. Joel Feinberg, "The Rights of Animals and Unborn Generations", in *Rights, Justice, and the Bounds of Liberty* (Princeton: Princeton University Press, 1980), pp. 159–84.

7. The possibility of basic human needs being changed by genetic engineering, or dependence on earthly resources being reduced by economic exploitation of space, cannot be ruled out in the very long run. But there will be many intervening generations dependent on earthly resources to supply their material needs, in any case.

8. Congressman have the advantage of being able to monitor some of the effects of policies on their constituents and modify the policies accordingly. Representatives of future generations may gain further *evidence* about the likely effects of policies on future generations, but will not be able to directly observe the effects. This is one reason that representing future people would be more difficult than representing present people.

9. Assuming that they will exist. Of course, we need not be *certain* that they will exist, to have reason to act to promote their interests, anymore than a bachelor has to be certain of living to retirement in order to have reason to contribute to his pension fund. On the question of whether we have moral reasons for seeing to it that future people exist, see Kavka, "The Futurity Problem", sec. 4, and Jonathan Bennett, "On Maximizing Happiness", secs. 5–10, in *Future Generations,* ed. Sikora and Barry.

10. Since other adults besides the parents (e.g., other members of extended families) can represent a child's interests, the actual situation is more complicated. But the net result is the same: some children have more proxy votes than others.

11. The possibility that future generations may not exist is often cited as a reason for ignoring their interests; perhaps it is not uncertainty, but rather a lack of imagination, that makes us care so little. The task of making the future live in our imaginations falls to artists as well as to scientists and public officials.

12. It is entirely rational to discount expected future benefits in virtue of the uncertainty of their receipt, or of one's later need for them, or of one's still being in a position to enjoy them. Discounting beyond this, absent further considerations, would constitute having an irrational "pure time preference".

13. Mary B. Williams, "Discounting Versus Maximum Sustainable Yield", in *Future Generations* ed. Sikora and Barry, pp. 169–79. One way of making indirect representation of future people more effective would be to change these faulty discounting methods.

14. Aristotle *Nicomachean Ethics* I. 10–11.

15. In Aristotle *Nicomachean Ethics* I. 10. 1101b. 1–5, this point seems to be acknowledged.

16. For this reason, we limit our proposal about direct political representation to apply only to future generations of the citizens of the nation in question. Since, however, many issues (e.g., nuclear-waste disposal) affecting the well-being of future citizens will similarly affect future foreigners, future non-citizens would receive a substantial degree of indirect representation via the direct representation of future citizens.

17. Realistically, no nation would seriously consider giving foreigners direct representation, unless other nations agreed to do likewise.

18. There is at least one exception. If the public intensely desired the protection of the interests of future generations, then having (and advertising) such a policy might increase sales.

19. Even if such groups represented future people only, this would not weaken the case for representation for future people within the government. After all, the existence of farmer's lobbies, for example, does not substitute for farmers having legislative representation.

On the Survival of Humankind
JAN NARVESON

The question to which I am to address myself herein is "Is there a moral obligation to promote species survival through the implementing of a eugenics programme? Does this obligation somehow involve consideration of the welfare of future generations?" There are, in fact, three different questions here, it seems to me. These are:
1. Is there a moral obligation to promote species survival?
2. If so, is it due to considerations about the welfare of future generations?
3. If there is that obligation, then does it carry with it an obligation to implement eugenics programmes in order to help carry it out?

The trouble is, I think that the answer to the *first* question is negative: that there is no obligation to promote species survival, at least there is no basic obligation of that kind. Curiously, the main reason that I think this is that I believe the answer to the second question is in the affirmative: *if* there were such an obligation, it would be because of considerations for the welfare, in some general sense, of future generations. That, as we shall see, is exactly why I don't think that the answer to the first question is yes. But if we move to the third, then I'm inclined to agree, hypothetically, that if there were such an obligation, then it at least could involve an obligation to use eugenics programmes in order to promote it. Insofar as there is that kind of an obligation, it might well lead to the obligation — or at very least the right — to use eugenics programmes. Since this is all no doubt a bit confusing under the circumstances, we had better proceed piecemeal and address ourselves to these component questions in their turn.

I
Species Survival and Obligation

Why might it be thought that we have an obligation to promote *species* survival? It might be useful, in considering this question, to make a couple of distinctions. When I refrain from assassinating my fellow humans, as I, like most of us, am in the habit of doing, there is a sort of residual sense in which I might be said to be "promoting the survival of the species"; but this is not what we have in mind in the present question. The *survival* of a given species (S) at a given time (t) is simply the existence of members of S at t, where t is a time later than some time (t') at which there already were such members. This obtains no matter which individual members of S there are at t. Thus assassins back at t' who are concerned to promote the survival of the species can, as far as their unsavoury trade is concerned, ply it at will provided only that they don't overdo it and wipe out either (a) the entire current membership of the species, or (b) all current members capable of reproducing successfully. (It is for this reason, incidentally, that opponents of abortion get no help from positing an obligation to perpetuate the species.)[1]

The presence of this variable "t" in the formulation above poses what could be a rather important question, namely, just which range of future times are we concerned with? Is the suggestion that each generation has an obligation to promote the survival of the species to the next generation? Or is it that each should take the longest possible view, and aim at the *indefinite* survival of the species? This would mean doing whatever was necessary to ensure that there be no last generation of the species – *no* time t such that there is no later time t' at which we could have seen to it that there be at least one member surviving and didn't? (Adding the last clause is essential, for presumably it could be that the end of the habitable portion of the universe for the species in question, owing to cosmic circumstances quite beyond our control, is foreseeable. In that event, it will not have been any culpable violation of our duty which made it the case that some generation was the last generation.) I believe that an obligation to promote the survival of the species would be in this form, that is, the obligation to promote the perpetuation, the indefinite survival of that species.

There are other conceivable goals which might be confused with that of promoting the survival of the species. In particular, suppose that someone thought that our obligation was, so to speak, to maximize the species, where by "maximizing" is meant "bringing it about that more members of that species have existed than any alternative policy would bring about". It is conceivable that the population of the universe in respect of that species (S) over time, is maximized by concentrating the bulk of it in the nearer future rather than by stringing it out, as it were, over a longer period. This would obtain if, for instance, there were some exhaustible resource (R) which was subject to two different exhaustion processes, one of them due to physical circumstances beyond our control, and the other due to the fact that members of S use it up as they live. It is also conceivable that the amount of R available at time t is such that fifty billion members could then exist, although this will cause the total depletion of R by t'; whereas if we were to cut the population at t to only, say, five billion, then there would still be some R around at t'. Indeed, conceivably the exhaustion curves on R are such that if we limit the population to, say, only a few dozen, then the last surviving member of S would live at a time t' a thousand years later than t! In these circumstances, those who opted for the fifty billion population at t could hardly be said to be aiming essentially at the *survival* of the species. They would instead be aiming at maximizing its total membership over time. The universe will have contained, when all the chips are in, a lot more members of S under their policy than under the alternative policy. Let us call these two goals, then, "perpetuation" and "maximization" respectively.

One interesting question to ask about these two policies is which, if either, makes the more sense? Those who think a certain way about these matters can be asked to consider some sizable fraction of the set of possible people who do not exist, given the above circumstances, if we pursue perpetuation rather than maximization. Someone who thought, for instance, that the existence of people (or whatever species is in question) was an *intrinsically good thing* ought surely to think it odd that the existence of the very few people in the very far future who would exist under perpetuation should be, just by the very fact that their dates are so much later, a better thing than the existence of so many people at the much earlier date!

To explain: the members of species all live in time – they all live *at* times. It is logically possible, then, to make a sort of God's eye enumeration of that species in the universe, not merely at a particular time but during all the time there is (or at least, all the time we want to consider!). Each member of S will have a pair of dates attached to it: the date of its birth (or conception, if you prefer) and its death. A member may be said to exist in the "timeless" sense of the term "exist" (though not, of course, to "exist timelessly" – that is not in the cards for members of the species we have in mind)[2] if the entire history of the universe (future as well as past) contains such a pair of dates for it.

Now from the selfish point of view of each of us, it matters a great deal which stretch of time some other people occupy. We want it to be the case that the stretch they occupy approximately coincides with our own (and in some other cases, we might wish that they had occupied a quite different, nonoverlapping stretch than we do!). But it is a little difficult to see how such factors could operate when we are contemplating alternative policies in relation to the rival goals of perpetuation and maximization. Anyone who would want to favor either of these policies in the abstract must surely be doing so on some sort of "intrinsic" grounds. If, for example, "S-hood" stands for "the property of being a member of S", then such a person presumably thinks that S-hood is an intrinsically good thing. And its being "intrinsically" good presumably means that it is so, not in relation to particular interests or values of the person whose view is in question here, but rather, somehow, good in itself. And whatever candidates there might be for the office of "intrinsic-good indicator", surely location of time cannot be among them. If we think that Epictetus and Hume were admirable men, for instance, and get into a discussion of which was the more admirable, surely the sheer fact that Hume lived about twenty centuries later than Epictetus is irrelevant to this discussion, whatever might be relevant!

But, on the other hand, it does seem plausible to suppose that if it is intrinsically good that P and it is intrinsically good that Q, then it would be intrinsically better still that both P and Q. Two intrinsically good things are better than one, are they not? (Possibly the two things cannot both exist. The premise that both of them should exist, if they could, is an academic matter — but this doesn't keep it from being true!) And anyone who believes this to be so ought to think that maximization is preferable to perpetuation, if a choice had to be made, other things being equal.

Then again, one might argue that populations strung out in time have one advantage over those all existing simultaneously. For the later members, with any luck, will inherit the knowledge and the culture of their predecessors. On both fronts, we could fairly reasonably hope that future generations will be better off than we are. Their lives will, in those respects, be more interesting. We might say, to put it somewhat fancifully, that if we had our choice between existing at time t and existing at t', where t' is much later than t, then, provided that other things were reasonably equal, we could reasonably prefer to live at t'.

Now, the "other things being reasonably equal" clause is very important in any such speculation. Suppose that if we lived at t', we would have a new and peculiarly horrible disease, unheard of at t and incurable at t', and that that disease would leave us bereft of any ability to absorb any of the knowledge and culture, greater amounts of which are no doubt there to be absorbed? Even if it didn't do that, suppose it just made life virtually unbearable, so that one could not enjoy the knowledge and culture in question even though one could acquire it.

This, of course, takes us into the question of the grounds on which one might advocate policies designed either to perpetuate or to maximize. And before getting further into those, let us simply address ourselves to the question of whether the existence of species, as such, can be regarded as intrinsic good or as something which we have some kind of basic obligation to promote. In doing so, I suppose, we had better advert to a question which I have tabled so far, namely: does it matter *which* species S is?

It is pertinent to consider for a moment the more general issue of whether species ought to be perpetuated, no matter which they are. Indeed, I think it has been held by some philosophers that a properly run universe ought to have, in principle, as many distinct species as possible. They are, so to speak, species maximizers in a new sense of the term from that defined above; for what they favour is precisely not that membership in any one species be maximal in the long run, but rather that there be as many species in the long run as there is, so to speak, room for in the universe. (I am not sure that this characterization quite does justice to their views, for perhaps they were in favour of as many species as possible existing simultaneously *and* as long as possible. Things get complicated at this point, for we can imagine that in the long run there would be more species if each of them only lasted so long; thus it might be necessary for a universe maker to take his or her choice between more distinct species each existing for a shorter period and less distinct species each existing for a longer, even perhaps a perpetual period. . . .) This idea, of course, would give only a rather modest weight to the goal of perpetuating our own species. For perhaps there are many species whose existence is incompatible with the existence of human beings, at least for any considerable period. Ignoring the possibility that one of those species incompatible with our own existence is *Homo sapiens,* it does seem more than a science-fiction possibility that there are some species which will have to go if human beings are to stay. And of course, if maximization is envisaged, the number of such species may be quite large. Perhaps all of the edible species of animals would have to go, and be replaced by more efficient vegetable species! Apart from that, let us consider such species as the black fly, or the mosquito. No doubt both of these have their roles in the economy of nature, to be ignored only at our ultimate peril. But if we address ourselves to the hypothetical question of whether there ought, other things being equal, to be black flies, I think we could be forgiven for coming down in the negative. Perhaps the black fly is a species we cannot do without, but it is one which a lot of us would rather do without if we could!

Now perhaps this is just one more of those "selfish interests" which I perhaps cavalierly dismissed above as relevant considerations in these contexts. But the problem is, why should the interests of those who profess a great interest in the maximization or perpetuation of species, just as such, be counted and our more palpable interest in the non-existence of the black fly not be? There is no doubt that some people have the interest in question, to be sure: we may call it the "collector's" interest. Just as there are some who collect postage stamps, so there are some who collect species. And we can well imagine that some collectors have interests which collide with those of others. If so, there is the question of why those with the species-collection interest should be given the veto — or even two votes — over those without it? And then, why should the species Homo sapiens, as it were, be given the veto, or even two votes, in comparison with any alternative species if it should come to a showdown? Suppose we could have several thousand new species if only it weren't for the existence of human beings. Why should we win, and the others lose?

Reflection on these questions will, I suspect, lead us to conclude that there isn't any really fundamental reason why there should be species of one sort or another, or more species rather than less, and so forth. But if this is true, then why should we think that the perpetuation of the human race in particular is an important moral goal, *as such*? For what we perpetuate is not, after all, the existence of any particular person — that is another question altogether! (Indeed, if we found ways to lengthen greatly the life span of individual people, we might well have to cut down drastically on the number of new people being born.) What we perpetuate under the heading of survival of the species is, then, the *species*. But if there is nothing to be said for the survival of species in general then is there anything to be said for the survival of *our* species merely as such? As far as I can see, there simply is not.

Future Welfare and Survival

The question we began with was whether there is an *obligation* to promote the survival of the species. But so far, I have considered only the question of whether the survival of species is an intrinsic good, or an interest of a kind relevant to morality. This raises the question of how obligation is related to such matters. Some theorists, after all, would have it that even if the survival of the species were an intrinsic good, it still wouldn't follow that we have an obligation to promote it. Might there be some other ground, unrelated to interests or intrinsic goods altogether? This is a rather deep question, but also one we can hardly discuss at length here. I shall have to rest content at this point with the observation that I have seen as yet no compelling reason for thinking so. But even within the realm of theories which have it that there can be no obligation where there is no interest or good in question at all, there are major and interesting divergences, and I wish to address myself to some of these. In brief, I find three theories worth considering. I will begin by stating them very briefly, then discussing the implications of two of them for the first question, which can be done very briefly indeed, and finally devoting rather more attention to the third.

All moral theories are concerned to guide the actions of moral agents. One such theory, however, has the distinction of proposing to guide it entirely on the basis of the *agent's* interests: whoever A might be, what A should do is whatever maximizes A's long-run satisfactions. Now morality is concerned with what we may call the interpersonal portion of practice, and what this theory will say about it is that morality is a sort of agreement among agents; in the absence of such agreements, things will go worse for each than if they are made and kept. A quite different view, at first sight, has it that agents have *rights* to act, to live their lives as they see fit, but how they may act and live is restricted by the similar right of all others. This second type of theory is "libertarianism".

Neither contractualism nor libertarianism give us basic obligations to perpetuate the human species. Contractualism does not, because we can, in this view, only have obligations to those whose actions can in turn affect us; and people in the distant future cannot. By the time they are on the scene, we have left it. If we don't want to worry about them, we don't have to. Libertarianism does not because in this view, people have only *negative rights*: we have the absolute duty to refrain from harming others — from interfering with the actions they have the right to perform. But we have no duty to assist them in performing them. Of course we have no duty to bring them into existence either. If we don't want to have any more humans around, as far as our participation in the project of having them is concerned, then we don't have to. Under these theories, then, if there is any moral interest in the perpetuation of humankind as a species, it would have to be indirect. The survival of the species is not, as such, something morality demands that we support or promote.

This brings us to the "utilitarian" view, which certainly seems to be coming out rather differently on this matter. For according to it, if an action (X) performed by some agent (A) would produce a benefit for anyone, then A ought to do X. We are to maximize the sum of benefit to all affectable parties. The parties in question might include nonhumans in this view, but we shall ignore that for present purposes. The question is whether this general goal of maximizing the sum of human happiness leads to an obligation of the kind we are investigating here. And here, as it turns out, the situation is unclear. What obscures it is this: in saying that we ought to benefit all who are capable of benefiting from our actions, maximizing the sum of those benefits and counting a given amount of benefit for any party equally with a given amount for any other, are we saying that we ought to benefit all the parties there *are*? Or are we instead saying that we ought to create, so to speak, all the benefit we can, even in the case where we have to create the *party* whose benefit is in question as well as the *benefit* in question? On the face of it, at least, there would seem to be quite a divergence between these two ways of looking at the matter. Given the first way, there would seem to be, again, simply no basic obligation to perpetuate the species. Population policy would have to be based on considerations of the benefit of whatever people there are, and of whatever people we decided there would be, but it would not involve any inherent bias favouring more people. Given the second way, there is a possible obligation of that kind. For it seems that on that understanding, we should keep adding people as long as the marginal benefit to the sum of human happiness from such additions is positive.

Perhaps it is not immediately obvious that there seems to be this difference between the two views about utilitarianism. It is worth dwelling on the matter further, especially since its interest goes beyond that of the utilitarian theory anyway. Let us, then, take the case where you are considering having a child. Suppose you are convinced that you ought to promote human happiness. And suppose you are also convinced that if you had a child, then that child would be quite happy. So far, so good. But now, do you have an obligation to produce the child? That depends. For we must now raise the question: on account of whose benefit should you do this? The interesting answer for this purpose is: "the child's". In producing a child who would be happy, you might suppose that it is for that child's benefit that you are doing it. But is this so? Surely not! For suppose you don't have the child. Can we now say that you have failed to benefit him or her? But the trouble is, people do not *exist* unless they are produced. So if you don't produce this child, then there is not only no benefit bestowed on it; there isn't even any question of you *not* having bestowed this benefit *on it*. The subject of this benefit or lack of benefit simply isn't there at all. It isn't, indeed, anywhere in the universe, past, present or future. So in the event that you don't produce a child at all, it is not even true that you have failed to promote the sum of human happiness. Not, that is to say, unless you mean that you have failed to bring it about that there are more instances of human happiness in the universe than there would otherwise have been. But that's the other theory, or rather, the other view of this theory. On one view, our obligation to promote welfare is always the obligation to promote *somebody's* welfare, and that seems to disallow us to count voluntary childlessness as a case of nonperformance of that duty. On the other view, we are to promote welfare even if there is as yet nobody whose welfare it is.

Which of these two views about utilitarianism is right is a question which has puzzled me for a long time. I am, indeed, still puzzled about it. And I am inclined to think that its resolution is a very deep matter, raising the most basic questions about the foundations of morality. But these difficult matters cannot occupy our attention here. In part I shall avoid them by being hypothetical and considering all possibilities; in part, I shall simply admit my bias in favor of the first way of regarding it, the one which does not seem to give rise to a direct duty to perpetuate the species. But it is important to note that even in the second view, this duty is not exactly "direct" either. What is direct is that we should maximize happiness, not *people*. The two need not, to put it mildly, be the same. It is perfectly possible that the sum of human happiness would be greater if there were less people rather than more. And it is possible that the sum of human happiness would be greater if the last living person lived much nearer to us in the future than would be the case on some alternative policy. "The more, the merrier" just isn't an implication of this view, not even if we add "other things being equal".

Two more points about the utilitarian type of view, on either construal, should be noted here. The first is that in both views, the location of a person in time makes no difference to his or her moral status. If it is obligatory that we treat a person in such-and-such a way, then we ought to treat that person in that way whether he exists at locations in time in the very far future or right now. It may, of course, be very difficult to do the thing to someone living five hundred years from now. It may be so difficult that we can practically ignore the matter — other things may well overwhelm any practical concern from this source. But whether that is so is a separate question; the fundamental irrelevance of remoteness in time, as such, is there in either view. It is, incidentally, also present in the libertarian view. People, wherever and whenever they may be, have rights not to have certain things done to them, and so we must concern ourselves about any future persons as well as present ones.

The second point is considerably trickier. There is a standard distinction in the philosophical profession between two sorts of duties: "strict", on the one hand, and 'supererogatory" on the other. The general idea is that a strict duty is one which you may properly be required to do — in particular, blamed for not doing, perhaps even punished for not doing; whereas a supererogatory duty is one which you should be praised for doing, but may not be blamed, or at least may not be properly punished, for not doing. Actually, there is even some question whether the supererogatory sort should be called "duties" at all. The tendency is to use "duty" and "strict duty" synonymously. Now, this wouldn't matter very much if it did not lead to a tendency to make out utilitarianism to be a very much more demanding doctrine than it may be. To illustrate, suppose that the difference between A's performing action X and A's performing action Y is only that X will produce one more lick of vanilla ice-cream by one person who enjoys vanilla ice-cream than would Y. It seems quite a bit to swallow (shall we say) to hold that in these circumstances, A could reasonably be *punished* for choosing Y instead of X! But there seems rather a promising way out of this unintuitive consequence for utilitarians. For they could urge that the disutility produced by punishing someone (let alone what is involved in the administrative overhead, so to speak, required in order to inflict a punishment) is much too great to be justified in so trivial a matter. Indeed, it is not worth producing even a very modest amount of moral discomfort for so trivial a result, we may well think. Instead, we should reserve such categories as the strongly obligatory and the punishable for much bigger game.

Reflection on this point may take us some considerable part of the way towards a practical result similar to that of one of the other theories, namely, libertarianism (though it could hardly take us all the way). For it might even be suggested that such things as blame and punishment should in general be reserved for acts which are positively harmful to others, instead of extending them also to acts which merely fail to benefit others. It is not implausible to think that the gain to humankind from this restriction would outweigh the benefits. Benefits bought at the price of coercing those who produce them are, it might be thought, benefits not worth the price. This was, indeed, the doctrine of John Stuart Mill himself, as we know; for Mill's famous "principle of liberty" explicitly enjoins us not to punish people for any actions except those which bring about "assignable harms to others".[3] The considerable plausibility of this idea also goes in the direction of reducing any pressure we might be inclined to put on those who don't help along the project of perpetuating the human race: the gains in utility from doing so, if any, might be cancelled by the costs involved in requiring those who are strongly disinclined to help to do so. Adding people to the population may well be a kind of action which should be left in the realm of things which people are free to do or not do as they please.

Thus far, then, my conclusion on the matter before us is that of the three moral theories I have any professional sympathy for, two of them simply don't support a duty to perpetute the human race at all, at least directly, while the remaining one has two interpretations, one of which also doesn't support it directly, and the other of which supports it at most in the limited sense that if the future people whom there would be if we perpetuate the human race make for a happier collection of humans on the whole than the collection there would be if we don't, then we ought to perpetuate it. And even this conclusion is much qualified by the reflection that it is hardly plausible to posit this outcome unless those participating in promoting it liked it enough so that they would do it of their own free will anyway, without any sense of moral obligation! So we may say that if there is an obligation to promote species survival, it is due to an obligation to promote welfare "somehow": namely, an obligation to promote welfare even if one must do so by creating those whose welfare it is! And that is an obligation which I am strongly inclined to think does not exist.

Eugenics and Survival

I have thus far been concerned to argue that there is no obligation to perpetuate the race, stemming either from welfare or any other considerations to which I find reason to attach any weight. No *moral* obligation, that is: or more precisely, no *fundamental* moral obligation. But I now wish to suggest that this isn't really the main consideration in this matter anyway. For, without any hint of moral obligation to keep the human race going, a great many of us surely have an interest in doing this. And without any shred of fundamental moral obligation to do things along that line, we might nevertheless have derivative ones from time to time. In both cases, moral questions can certainly still arise. In the first place, they quite definitely arise in regard to the *ways* in which we might try to promote this interest, which so many of us do have. We want to know what moral restrictions operate in this area. And regarding the various derivative obligations we might incur, there are the questions both of what they might obligate us to do in particular and, again, of how they relate to possibly conflicting obligations.

To make these remarks a little less abstract, I will amplify a bit. I have suggested that many of us have "an interest in keeping the human race going". Actually, we have two distinguishable interests here. Firstly, of course, it is an obvious fact about most people that they *like* having children. It is true that many of them also feel an obligation to have them; and of course many no doubt have families because it is the custom, or out of unconsidered habit. But I am enough of an optimist (if that is the right word) to think that most people would want to have families even if they considered their hitherto unconsidered habits, and even if they persuaded themselves, as I have myself, that obligation has nothing to do with it. Now this is not the same thing as an interest in perpetuating the race, as such. It does, however, have much the same effect. It is a little like the "invisible hand" processes which have fascinated some philosophers: each contributing person's intention is merely to have a child or two or three, but of course when those children grow up and come to have the same intentions, and their children in turn, etc., the race is perpetuated willy-nilly. In addition, though, there is a second, separate interest in perpetuating humankind, which I am quite sure many people hold. These people may, for our purposes, be divided into two classes. Firstly, there are those who have a variety of ulterior motives for doing so, such as a desire for post-humous fame (a desire which requires for its satisfaction that there be a posterity to have the requisite admiration); or an interest in having "higher types" around, which is seen to require a lengthy process of natural (or unnatural) selection; or a desire that a great deal more be known about the universe than is known right now, which is seen to require many more generations of scientific progress and hence of scientists to promote it. But secondly, there are some who are "collectors" in the sense discussed earlier this paper — people who simply want it to be the case that some specimens of *Homo sapiens* be around at any given time, if possible. It's a matter of giving the place tone, so to speak: no self-respecting universe, they think, should be without at least a few humans.

Now, as to derivative obligations, it seems to me that there are two sorts of interest here. In the first place, of course, people often take on the obligation to have children. Their spouses, for instance, insist on it, or perhaps they promise to supply grandchildren to their parents. Secondly, we may sometimes have the view that in order for the people who presently occupy a particular area to flourish, there really needs to be more of them. Now if, as utilitarians think, we have an obligation to promote the wellbeing of humankind, then this might generate an obligation to support population growth. I have heard supporters of the Canadian "baby bonus" argue for it in precisely these terms; and whether or not the argument was a sound one, the principle is quite coherent, I believe.

So we come now to the question of how all this bears on the use of eugenics. In my view, the main question is not about the obligatoriness of their use, but rather about the legitimacy of such methods. That, at any rate, is what I shall mainly consider. But before doing that, it should be noted that eugenics, as that term is normally used, is essentially directed towards improving the species rather than towards perpetuating it. Under what circumstances might it be required for the latter rather than the more usual former goal?

In reflecting on this matter, some rather intriguing possibilities suggest themselves. For example, it is not beyond possibility (is it?) that these two goals, at least as they are usually conceived, should come into conflict. Take, for instance, the improvement of human intelligence. Without getting into the recently politicized issue of whether there is an identifiable trait called "intelligence" and whether it is literally inherited (that is, genetically), let us simply suppose both. But let us suppose that we also become convinced (not entirely unreasonably, I might add) that human intelligence is actually self-destructive on a global level. Various scenarios can be envisaged: some intelligent people are malevolent even now, and perhaps we discover that this property is inheritable and yet unpredictable, a certain random percentage of all intelligent people being fated to be Dr Moriartys or Dr Strangeloves. And as knowledge grows so — as we have very good reason to believe — grows the knowledge of how to do people in. Perhaps the time comes when we can foresee that the probability of someone's coming along who will render the species extinct gets uncomfortably close to 1.0 *unless* we undertake a rigorous programme of, well, not exactly *eugenics,* but a programme designed to weed out all members of the species with I.Q.s in excess of, say, 140. We can keep the normally and only moderately abnormally intelligent from blowing us all to bits, but the geniuses, alas, will have to go. What *then*? A painful choice, surely: not merely because geniuses are intrinsically terrific and lead terribly interesting lives which the rest of us envy, but also because they're so much fun to have around, not to mention so useful (always inventing things, such as the light bulb and Beethoven's Missa Solemnis, and so forth)! Well, under these sad circumstances, would we have the *right* to insist that every potential parent of a genius be sterilized, in the interests of the survival of the species?

In this case, incidentally, there would also be the question of whether we have the duty to do so. For even if there is no duty to perpetuate the human race, there is surely a duty to protect the current and future members from obliteration, isn't there? But even so, there is also the little matter of the right to have offspring, and any number of other rights people are inclined to credit us with which would probably have to be invaded thoroughly in order to make such a programme work. Are these costs that people ought to be willing to pay in the interests of the survival of the species? Are they costs that we have the right to *make* them pay if they don't want to?

I'm sure that most of us believe that we won't have to make the sort of choice described above. Either there will be ways of containing the genocidal tendencies of geniuses (maybe some other genius will discover the ways!) or the posited connection simply doesn't exist (we'll all get blown to shreds by some idiot instead, or by some perfectly normal person in a fit of pique). So let us address ourselves instead to two much narrower questions. Let us suppose that there isn't this incompatibility at all, but that instead we discovered an inheritable trait of malevolence, as such, that we could predict precisely which parents would produce a person of that sort, and that only such persons were likely to cause the premature termination of humankind. In the extreme case, indeed, we would know not only precisely which offspring would be *likely* to do such a thing, but also precisely which offspring would actually *do* it. (A major hitch in the latter fantasy is that if we knew that, we could surely prevent him from doing it even if he were born.) Well, don't we all think that in this extreme and highly fanciful situation, we would have the right to prevent those parents from producing that child, even against their will? That preventive measure, given the situation as described, would seem to be entailed by the right of self-defense.

This brings us to the second case, which differs from the first only in the respect that, instead of knowing exactly which individual, if produced, would exterminate humankind, we instead know which sizable class of parents is such that there is a small probability that their children would do that. The probability might be very small: one in a million, say. Well, is that a high enough probability of a thing like that to justify us in forcibly sterilizing that *entire set* of potential parents? And, if not, why not? After all, the lives of billions are at stake; and we are not, after all, posing the standard and more oppressive question people are always raising for utilitarians, viz., whether we may legitimately kill a million in order to save the lives of a billion. Instead, we are merely asking whether we can sterilize a million in order to save several billions. (As I have noted above, you do not deprive a potential person of life by refraining from having a child when you could. Merely potential persons cannot be "deprived" of anything.)

It is interesting to note that this is an issue which cannot be easily resolved by using the principles and conceptual machinery of libertarianism, as some might suppose. It would clearly not be possible, for example, to rule that the act of exterminating humankind is one which we may not forbid but, instead, can take care of by allowing it and exacting suitable compensation! It seems that what we must do is to determine the price that we should have to pay those who were involuntarily sterilized given the probability that their offspring would prove to be genocidal maniacs. This price would be less than the amount we would have to pay someone whose children would be normal not to have those children, but it could not easily be nothing at all. I believe the libertarian view on this matter is correct. We cannot go sterilizing people without their consent for anything less than some very high probability that their children would be exceedingly dangerous (even more dangerous than ordinary people are, for instance!) or exceedingly miserable. But where does the right amount lie between these extremes? Clearly, it is very difficult to say. Perhaps the relevant variable here is the amount it would cost to contain the propensities of these dangerous offspring if they were born. If this amount is collectible if the potential parents in question go ahead, then presumably we could deduct it from the amount due to them in compensation for depriving them of the ability to have children. Rather high finance, perhaps, but it offers some kind of hold on this problem in principle, at least. (Such people, if there are any, would also do well to re-educate themselves on the merits of adopted children, to which there would be no similar objection.)

Short of genetic propensities to race extermination, what other situations might raise the question of using eugenic methods with a view to ensuring the survival of the human species? Here is one which brings up rather different issues than the foregoing. Let us suppose that some inordinately attractive group of people arises, who carry a genetic factor which somehow brings about sterility in, say, their great-grandchildren. (I don't know whether this is genetically possible, but the question here is strictly academic.) In the preceding couple of generations, however, they are, in addition to being inordinately attractive, quite extraordinarily fertile. Conceivably, every normal potential parent could end up mating one of these people, in which case the world would suffer genetically induced depopulation in about a hundred years. If the original group were readily recognizable, would even this peculiar eventuality legitimize compulsory eugenic methods?

I think not. If they are readily recognizable, then people who have a strong interest in the survival of humankind will avoid reproducing with these people. Even if one succumbed to the inordinate attractions of one of these inordinately attractive people, he or she could arrange the reproductive side of their relationship in some other manner. Indeed, there is no reason in principle why one of these special people should not agree that the survival of the species is a good thing, and so agree that the reproductive cells by means of which their children were conceived came from normal sources — via a sperm bank, say, or an egg transplant.

There may be people who think that isn't good enough. If so, we may pose to them the following question. Let us suppose that instead of the genetic situation envisaged above, what happens is that a group of exceedingly persuasive people arises, who persuade people that the human race is not worth carrying on, or that only total abstention from sexual activities is compatible with true purity of soul (plus, of course, the suppressed major premise that purity of soul is a really good thing to have). Now you, a member of a small band of not very persuasive but biologically resourceful persons with a powerful attachment to the survival of the species, dream up a method of rigging people's genetic make-up in such a way that their offspring are, in effect, genetically inoculated against the persuasiveness of the persuasive group. Would you be justified in subtly contaminating their water supply so that the persuaders will be foiled from the outset?

It seems obvious to me, I must say, that you would not. There are at least two reasons for this. One is that if your side reproduces and theirs does not, it seems obvious that you have a strategy available which guarantees that proponents of your view will prevail in the long run. And in the second place, what if, contrary to your present desires, you yourself get persuaded by these highly persuasive people? You will then have ensured that a great many people will exist who have no chance of attaining true purity of soul. Not to mention that the method proposed would involve a horrendous violation of intellectual liberty.

All in all, I find it difficult, as will be evident by now, to envisage circumstances in which eugenic methods would be necessary to ensure the survival of the species. And in those where they would, it is by no means a foregone conclusion that their use would be justified. And I wonder whether people who hastily advocate such procedures aren't suffering from a curious sort of paranoia, or at least neurotic insecurity. We are sure that the survival of the species, at least as long as it is on reasonably good terms for those doing the surviving, is a good thing. And it is tolerably obvious that most people think the same way. Since it is also tolerably obvious how one goes about promoting that goal, and that the free activities of those who share the belief is quite sufficient to ensure it, what kind of sense is there in supposing that in addition to these free activities, we must also force all those who happen not to share the view to act as if they did? Under no reasonable moral theory would that be justified.

Notes

1. Cf. Jan Narveson, "Semantics, Future Generations and the Abortion Problem", *Social Theory and Practice* (Fall 1975): 473–74.
2. See Jan Narveson, "Utilitarianism and New Generations", *Mind* (Jan. 1967), idem, "Moral Problems of Population", *The Monist* (Jan. 1973), and idem, "Future People and Us", in *Obligations to Future Generations* ed. Richard Sikora and Brian Barry (Philadelphia: Temple University Press, 1978).
3. John Stuart Mill's *On Liberty* has the main argument. See the very penetrating discussion of this in D. G. Brown, "Mill on Liberty and Morality", *Philosophical Review* (April 1972).

On Deep Versus Shallow Theories of Environmental Pollution

C.A. HOOKER

"... 'pollution', with its twin suggestions of sacrilege and impurity. These emotional overtones ... increase the danger that anti-pollution measures will be overdone, that men will pay more for the reduction of pollution than, on rational consideration, they should have wished to do."

"In so far as the ecological movement necessarily involves a critique of the creeping bureaucratization so characteristic of our society, it has a social and political importance extending well beyond its immediate ecological ends; in so far as its practical policies would tend to encourage that bureaucratisation, it represents a social danger. For not only are man and nature fragile, so too is liberal civilisation, as the tragic history of inter-war Germany so vividly reminds us."

John Passmore*

What, after all, are the goals of philosophy?

I see philosophy as the attempt to construct a general view of the world: view = theory + commitment. The goal of the construction is understanding (as opposed, e.g., to mere knowledge) and, through understanding, the achievement of wisdom.

That to understand is to theorize follows from our evolutionary ignorance, requiring that we grasp the world through imaginative construction put forward in bold hope of fitness. Epistemology is evolutionary and naturalistic; more generally, philosophical theory proceeds in explicit dynamic interaction with scientific theory; there is no "first philosophy".[1]

From this point of view it is apt and fitting that philosophers should be deeply involved in the articulation and criticism of comprehensive

* John Passmore, *Man's Responsibility For Nature* (London: Duckworth, 1974) pp. 60, 183. Further references to this work noted within the text.

views concerning the nature and resolution of what are called our environmental problems. At the very least these problems pose interesting applied problems in ethical and social-choice theory. At best they offer the philosopher the opportunity to articulate the lineaments of an entire culture, its expression in terms of social institutions and practices, and the systematic consequences of these for environmental relations. The philosopher's role is much more than auxiliary analyst, it involves the articulation of a wise way of life and the diagnosis of the deep roots of unwisdom in extant cultures. (Appropriately, the analysis is not to be carried alliteratively/impressionistically but to include as rigorous as possible analysis of the component parts — for example, of collective decision making, institutional design, human psychology — in the spirit in which science and philsoophy are one.)

Of course there is always the irresponsible claim that philosophy is only conceptual clarification — as if *that* were possible without commitments; without, in fact, employing many substantive theories and presupposing a metaphilosophy.

In 1974 John Passmore published a book, *Man's Responsibility for Nature,* born out of concern for the environmental issues then at centre stage intellectually and politically. In what follows I propose to critically examine one major line of argument in the book, not least as a vehicle for expressing my own concerned analysis of environmental problems.[2]

In the book Passmore claims no more than to clarify, and sometimes talks as if he might espouse a metaphilosophical doctrine to the effect that philosophy is substantively neutral, seeking only therapeutic clarification. Fortunately he is too serious a philosopher to rest his work there, and his book contains a great deal of theoretical analysis and substantive commitment. Unfortunately (given my values) the suggestions of a noncommital clarificatory role disguise a deep-seated conservatism of intellect as well as of political persuasion. The flavour of this conservatism is indicated in the quotations with which I opened.

I
Interlude: Passmore's Philosophical Contribution

According to the account which I shall later offer, environmental problems emerge from the deepest structures of western life — physical, societal, cultural, personal. Understanding them and responding to them properly requires a theory and practice of the healthy society and the healthy individual within it, and a theory of environmental design; the three cohering to form a unified conception of the future of the planet. Response to environmental problems calls for a vision of life. Responding to this need is philosophizing in the deepest sense, it is to seek wisdom — but also to elaborate a systematic philosophy of the person, society and the ends of human life.

Passmore admittedly has a more modest aim, to examine only humanity's responsibility for nature. But this cannot be examined independently of the whole, nor does Passmore succeed; his book includes discussions of ethics, economics and a vast range of other topics. The book is set out as follows:

> "It begins with two chapters which are essentially historical, in which I set out to describe those Western traditions which tend to encourage and those which might serve to curb man's ecological destructiveness. Then I examine in turn, now in a manner which is analytical rather than historical, what are by general consent the four major ecological problems: pollution, the depletion of natural resources, the destruction of species, over-population. In the final chapter I set in confrontation Western traditions and ecological problems. I ask what the West has to jettison, and what to retain, if it is to have any prospect of solving the problems which confront it (p. ix).

In fact the book's content falls roughly into two categories: (i) philosophical analysis and evaluation of relevant ethical issues; and (ii) development of a framework for redressing environmental issues. The bulk of the book is devoted to the former sort of material: all of part 1 ("The Traditions" chapters 1 and 2), the bulk of chapters 4, 5 and 6 in part 2 ("Ecological Problems"), and much of part 3 ("The Traditions Reconsidered", chapter 7). The main development of a substantive theory for redressing environmental issues comes in chapter 3, on pollution, with fragments of practical advice scattered through the remaining chapters. I shall concentrate on this latter material (substantive theory) in this essay, but before doing so I want to include some brief remarks on Passmore's ethical analysis.

From my point of view, the best part of the book by far is part 1 concerned with analysing the historical themes in Western culture having to do with the relationship between humankind and nature. Passmore argues effectively that there is not one but several themes, that the recently much commented on Christian theme of "man as despot" is not the only Christian theme. In particular, Passmore isolates a "man-as-steward" theme, which is conservationist in outlook and a "man-as-creator" theme which is more boldly interventionist in attitude and looks to humanity as a perfecter of nature. Both themes, the former especially, recognize nature as existing to glorify God, not merely to serve humanity, and both lead to a view of nature as both valuable in itself and valuable for humanity.

Moreover, I found that in chapter 4 Passmore made a valuable addition to my thinking about responsibilities to the future with his pleading the special demands of love; where market prudence and even "justice as fairness" will not get us very far in this context, the bonds of love enlarge the spirit and our commitments. This doctrine is more required in our unique historical circumstances where our decisions have exploded to encompass the planet — no simple inductive rationality, local or long term, can guide us adequately; only a vision of the human future which love bears provides a rich enough framework to assess action.[3]

I brush over the chapters on preservation and multiplication (5 and 6). There is much of value here in the balancing of opposing argument, the sensitivity to presupposition and nonsequitur. My immediate quibbles are few, my delight familiar. The gentle conservatism is there; whether (as I suspect) it ultimately undermines the air of reasonableness permeating the treatment, as I shall now argue it does for the treatment of pollution, I have not attempted to discover.

II
Passmore's Theory of Pollution

The chapter on pollution begins with a homely definition: "pollution is simply the process of putting matter in such a place [i.e. the wrong place — CAH] in quantities that are too large" (p. 45) and a homely remark about such problems being in general neither freely tolerable nor wholly soluble, a suitable compromise being called for. The homeliness is not only a measure of Passmore's "good sense" but also a measure of the problem; pollution is chosen first because it is a sufficiently simple problem that we may see in its treatment the lineaments of the treatment of other, more complex issues (e.g., those of succeeding chapters).

We are then presented with a model for the systematic treatment of pollution problems: (i) scientific investigation provides a thorough understanding of "how a particular form of pollution arises and in what its danger consists" (p. 48); (ii) technology then discovers "a method of reducing its incidence" (p. 48), in fact several alternative methods are usually available; indeed (iii) economists next provide a comparative table of costs and benefits for the various alternatives technology provides — at least they do this to the extent that the confines of their practice permit; and finally (iv) the preferred solution is examined for its moral acceptability, political feasibility and administrative consequences; after which (v) it, or a reworked proposal emerging at step iv, is implemented.

Along the way through these five steps there is some sane and insightful discussion. Cost-benefit analyses, Passmore reminds, may be confined because of ignorance, difficulties of quantification and the deeper difficulty of fundamental differences in evaluation. These are taken as warnings to scrutinize such analyses carefully, not to abandon them. Politically, Passmore emphasizes the larger role of culture and social practice as shapers of what is politically feasible, beyond mere self-interest. He is centrally concerned too with the way in which big bureaucracy, as much as big business, can hinder pollution control and, conversely, with the naive environmentalist appeal to a utopian central government that would in all likelihood only yield another totalitarian state. There is a wealth of sane consideration and advice in these pages, but is it a sufficient framework to understand and respond to our environment problems?

Consider for a moment Passmore's treatment of the moral issues. "Moral problems", he tells us, "do not hold up the solution of pollution problems, even if changes in our moral outlook might make them easier to solve" (p. 57). As the main grounds for this optimism Passmore cites the West's moral acceptance of state intervention to prevent damage of one party by another and the moral presumption that beauty, natural and artificial, should be preserved and (respectively) created. Building on this base, he suggests that societies will come to value "superior goods" more highly than their more-basic "inferior" counterparts (e.g., natural beauty over efficiency, low-waste production over cheapness) as they slowly grow more affluent. This gentle optimism reflects nothing of the anguished cries of the artist as he travels through streets made ugly by ill-design, unkempt buildings and exploitative advertising, nor that of the naturalist as he watches species die out and ecologies transform under wheel tracks, bulldozers and litter. It reflects nothing of the intensity of feeling, the identification with the cosmos as a whole, that drives the antichauvinist ethicist, the religious awe of life of a Jainist, a Teilhard de Chardin or a C. S. Lewis.[4] It is not that I want to especially advocate one of these views, but there is all the difference in the world between coming reluctantly to compromise from a deep-felt position on human values, and complacently "incrementing along", content with our sensible moral progress. The parting of the ways here concerns whether deep vision or piecemeal commonsense best reach for the Ture and the Good, and drive human development along.

Another indicator of the comfortableness from which Passmore speaks is his professed preference for "close communication among specialists" over reliance on "generalists" (his scare quotes) in approaching complex ecological problems (note p. 48). Granted specialists must be involved, why must this be the only right approach; professor to professor, as it were? My own experience in interdisciplinary investigations suggests that both types have their valuable roles. For example, some of the most valuable work done in energy policy over the past decade has been done by generalists who are reacting to and coordinating specialists and who perhaps have transcended their own specialist pasts. However, the optimally useful roles seem to vary significantly with the problem; I know of no safe rule except to be one's own specialist, and generalist, as much as possible. What is needed is not simplistic dichotomies based on hunches but some serious design studies of complex problem solving. Underlying design considerations, I shall argue, are crucial factors that are lost in Passmore's approach.

The absence of design becomes clearer in what Passmore *doesn't* say in his evaluation of cost-benefit analysis as a rational decision tool. He reminds us that such evaluations are often confined by ignorance, suffer from considerations not easily quantified and from deeply divergent views deeply held (pp. 50–53). Agreed, but what of those cases where these factors are at a minimum, or circumventable for the kinds of special reasons Passmore mentions (e.g., a known threshhold, obviating need for further details)? In that case Passmore happily endorses the procedure (p. 53). But this leaves what is to my mind the deepest feature of cost-benefit analysis in this context unexamined, namely, the relation of cost-benefit categories to global designs. Energy choices provide an excellent example.

Suppose a country has, at time t_o, a well-developed, coal-fired, thermal-electric power system serving a wide range of energy needs. Further suppose that increasing energy demands require additional generation technology to meet the demand and suppose that it is desired to evaluate the attractiveness of introducing solar-driven technologies against an expansion in coal/thermal generation for this purpose. Because of massive front-end capital costs and significant uncertainties associated with the solar option, an incremental expansion of the coal/thermal option has an overwhelming incremental advantage over the solar option. In fact the situation could be worsened considerably and the same result would obtain. For example, short of political expediency the same cost-benefit considerations would, in realistic cases as at 1981, permit massive increases in coal price and/or in medical compensation costs for coal-induced lung diseases before solar energy would be cost-competitive over, say, a five-year horizon. In fact these precisely are the arguments that have been largely used over the past decade to justify incremental expansions of conventional energy technologies as against investment in alternative technologies. The survival interests of entrenched energy bureaucracies only add to the "costs" of any alternative.

But suppose instead that, at time t_o -40 years, the same society had chosen to invest in solar technology. There would now be massive societal investment in solar industries, solar units would be made much more cheaply through mass production; long experience would have eliminated most of the investment uncertainties; decades of industry- and government-backed research would have greatly improved performance, safety and social convenience; solar-oriented bureaucracies (e.g., to inspect, tax and service) would identify their interests with solar expansion; and so on. Now imagine trying to introduce coal/thermal technology at t_o on a rational cost-benefit basis!

The fact is that incremental cost-benefit analysis, as practised by most economists, is more than a little like a discussion among lemmings on their suicidal race to the sea cliffs about whether they should stop to explore alternatives. Unless the ground is very rough or a deep transverse valley appears — unless an established technology is virtually falling apart or an overwhelming technological breakthrough is made — the gathered momentum of the charge carries it forward to near the brink of the precipice before there is any reason to re-evaluate the course. But by then it is too late to avoid the plunge. In a country such as Canada which has been investing in nuclear electric technology at a gathering momentum during the 1970s, the next "plunge" would have come around 2015—30 when, having undergone an economically exhausting and socially bruising round of nuclear investment in uranium "burner" plants, a new, much costlier round of investment in thorium-plutonium-burning technology would have had to be entered to offset declining uranium supplies.[5]

If one wants to stop the lemming charge, one has to provide alternatives *early on* and adopt an appropriate planning process to ensure their development — but incremental cost-benefit analysis is powerless to help in this decision; within limits it always reinforces existing commitments. In the energy-policy field the alternative approach that has emerged is *retrospective-path analysis*: a variety of end-points, for example, solar, nuclear and oil-coal technologies, specified socially as well as economically and technologically, are described and evaluated at an appropriate future time (say, forty to fifty years on) and by some social process a preferred mix of them chosen (with suitably hedged bets for unknowns); this end-goal is then worked retrospectively to decide the crucial branchpoints in societal decisions which will guide the market (working on an incremental basis) in the desired direction. Of course this is too simplistic, as described: there are often real technological and economic constraints on some transitions; ignorance demands that the decisions be constantly re-examined, etc. But the point remains that this approach provides a very different *design* setting for the application of cost-benefit analysis, it affects the determination of what count as costs and benefits, the approach to risk in the equation and perhaps even the rationality criteria themselves.[6]

Passmore, though clearly aware of the subtlety of the considerations typically involved (e.g., he mentions indexes of community suspicion while discussing the costs and benefits of life in various cities [pp. 50–54]) and though in another chapter it becomes clear that he is aware of the "lemming effect" (see pp. 85–86), yet he lets no hint past that there might be larger structural issues involved that lead to very different approaches to public policy making.

Another important structural issue is the definition of the problem to which cost-benefit analysis is to be applied. Consider the following fragment of a Martian report on our practices: "Industrialized humans take their most obvious naturally occurring waste nutrient (human and agricultural excreta) and, instead of using it to produce useful plant growth, dump it into the subecology most sensitive to it as pollution (rivers, lakes and sea shores). They then complain about the aesthetic, recreational and commercial results, the cost of repairing the ecological damage, the cost of agricultural fertilizers, the consequences of their flow into the same subecology, the potential long term agricultural costs of purely chemical growth promoters and the quality of the resultant foods!" Where there could be a structurally cyclical system, returning agricultural wastes as humus to the soil, there is a linear system: resource extraction, chemical factory, plant growth, consumption, water (as dump). During the North American phosphates scare (phosphates from farms, dishwashers, etc., were promoting algae growth and decay, thereby de-oxygenating the fresh waters) how to make an adequate detergent that did not lead to de-oxygenated water became an urgent problem. This was an urgent problem only within the context of the linear-system design; within the alternative cyclical design (or any one of a dozen others) the problem vanishes – in fact, in the cyclical design phosphates become a positive asset, they are an essential component of fertilizers.

At the height of the scare the introduction of a phosphate substitute, NTA, was seriously contemplated, despite suspicion that it may be a carcinogen. Its introduction was contemplated because, on cost-benefit grounds, the solution of the ecological problem was held to be worth the risk. From a larger system point of view, where the role of phosphates may be reversed, this is an excellent example of a *systematically misposed problem* – yet cost-benefit analysis can be applied to it quite happily once the larger system structure, which defines the problem, has been fixed. Systematically misposed problems can be dissolved as well as "solved", and often should be dissolved. Only an appropriate examination of the larger systematic context can reveal whether or not problems are systematically misposed, but Passmore gives no hint of the necessity of this critical framework for cost-benefit analysis (though he does recognize a larger context where problem solutions interact [cf. his discussion of phosphates at p. 66]). To my knowledge, only retrospective path analysis, with its systematic examination of alternative futures, has the ability to reveal systematic structural or design features and thus to alert one to potential misposed problems.

The politics associated with incremental cost-benefit and retrospective path analysis approaches are also very different. In the incremental approach it is the micro costs and benefits determined by inputs, outputs and risks to private individuals (whether persons, firms or specialized government corporations) that dominate the equation. The system structure taken as a whole is aggregated out of these individual decisions. In the retrospective analysis it is global systematic features of the outcome that must dominate the considerations because these will determine, or at least constrain, which kinds of private activity have favourable cost-benefit balances. Thus, the latter emphasizes global, systematic societal decisions rather than micro, individual private decisions and must accordingly place great importance on the political processes by which these decisions are made, processes which are assumed automatic (by aggregation) or merely facilitating in the incremental approach. I believe it to be a profound feature of our history that the scales of our decisions, spatial, temporal and institutional, have increased rapidly and are still increasing, so that responsible decision making (*a fortiori* rational decision making) calls increasingly for a global, systematic approach. In the light of this, I believe we urgently need sound advice on how humane, just and intelligent decision processes can survive the leaping scale of decisions and of the institutionalized concentration of power.[7]

On every side one sees institutional developments which are hardly paragons of these virtues. On the "left" dictatorial state-socialist/ communist bureaucracies stiffen their absolute power with self-righteous claims to the moral truth. On the "right" we find gigantic multinational corporations whose actions are dictated by short-term private interests but whose information, power and privacy exclude most participation and reduce social relations to those of employee or consumer, neither case offering much by way of autonomy, respect, growth, affection, etc., and equally large bureaucracies whose practical performance and social effect is all too often very similar (cf. Passmore, pp. 61–63). (Further still to the "right" lie those fascist regimes whose lumbering institutions and practice, if not their metaphysical rhetoric, are indistinguishable from severe central state communism.) Neither alternative attracts, each has its distinctive advantages and defects; but in my mind they all share an overwhelming disadvantage, for all, in their several ways, promote the creation of powerful elites in charge of huge bureaucracies which are able to manipulate, threaten and perhaps destroy those perceived to be at cross purposes to their goals. I find increasingly that it matters less to me whether one is coming from the "left" or the "right" as long as one is seeking more participatory solutions that will effectively distribute social power.

Social power needs to be spread within limits, otherwise the associated costs are too high. Anarchism will spread social power but at the cost of being unworkable in practice and/or surrendering much of the benefit modern life affords. In this ultimate balancing of "costs of passage" against "benefits of arrival" I am in Passmore's pragmatist spirit. Passmore too is against big bureaucracy and unfettered corporations (cf. pp. 61–63, 67), but he also speaks approvingly in the same pages of increasing state intervention (pp. 61, 66), and he fails to recognize that all of his canvassed, economic solutions for pollution abatement (compensation, rights, fines, taxes, [pp. 68–71]) involve the social equivalent of bureaucratic policing. Again there is much good sense in these pages, but theoretically I detect only an unstructured pragmatism. But I believe it is important to understand, as deeply as possible, from whence one is fleeing and toward which destination one is aiming. At any given time the practical upshot may share much in common with Passmore's gentle pragmatism, but the deeper differences are crucial and in the long term the two approaches can be counted on to diverge more and more. It is no easy matter to say how to spread social power without, on the one side, making life unworkable or, on the other side, producing unstable institutions that quickly assume a more extreme form.

The foregoing considerations concerning global design, policy-making theories, systemically misposed problems and political/institutional process theories don't permit much hope that Passmore's overall structure – science → technology → economics → morality/politics/administration – will prove either very insightful or helpful. It is naive to think that pollution problems come simply identified as such. Although Passmore is aware that they are evaluatively defined, not objective (p. 46), he offers no hint that they might be systematically misleading, needing to be dissolved rather than solved. It is equally naive to think that technological investigation comes after scientific investigation and before moral, political and administrative considerations. Many of the policy choices in the energy area say twenty years hence will be strongly influenced by technologies available then, by available information on their human and environmental inpacts and by whether or not supporting legal and administrative frameworks exist! However, the presence of these factors will in turn depend heavily on policy decisions made now to encourage research and investment in selected energy areas – and these are pre-eminently political and administrative decisions (yes, and moral decisions too). We are involved with, not a simple linear system operating serially, but a multiple interacting dynamic that can as easily back itself into self-determining blind alleys as show really intelligent and responsible behaviour. In view of the sensibilities and sensitivites Passmore displays it could be unfair to take a naive linear approach as representing his full intuitive appreciation of pollution problems; my complaint is rather that he fails to provide an illuminating and critical theory of environmental problems, which is the one contribution a philosopher might be expected to make.

III
Framework Sketch for a Theory of Environmental Problems

The following remarks constitute only an extremely crude sketch of a position, there is no space for filling out detail here (and anyway filling in some of the detail involves completing several fundamental research programmes). My purpose is to place the foregoing remarks critical of Passmore in a more systematic context.

> *Proposition 1:* Judgments of environmental problems are evaluative; they consist of evaluations of a state, process, etc., as undesirable, evaluations which may be backed by a variety of scientific and broadly cultural statements.

This is obvious on reflection (even devastated ecologies may be preferred by some), but it is of primary importance because it sends us looking for the value judgments behind all such assertions and in this way carries us immediately to such questions as: "Which sets of values, if any, are rationally preferable?", "What is it to have a philosophy of culture?", "Which culture(s), if any, are rationally preferable?" and other difficult but extremely important questions concerned with the fundamental structure of a coherent world view.

Practices that lead to environmental problems require careful examination of particular cultural histories and psychological dynamics, (some of which Passmore offers us). In the next proposition, however, I make a bold (and crude) attempt to grasp the core structure of these practices in Western industrialized nations through reference to abstractions from our cultural history and psychological dynamics.

> *Proposition 2:* In Western industrialized market societies, environmental problems appear as external diseconomies.

The economist's idea of an externality is roughly this: an effect (benefit or cost) which one action has on another's economic condition but for which the initiator is not economically accountable. For example: A is a market gardener whose run-off carries phosphates into Lake Macquarie affecting B's fishing business. If the affect is adverse B cannot (at present) sue A for damages; we say that A's farming creates an external diseconomy for B. If the effect is favourable B is not required (at present) to share his profits with A; we say that A's activity creates an external economy for B. For my purposes public goods (e.g., pleasant parks, well-designed cities, efficient administration) and public bads (the opposites of public goods) can be counted as external economies and diseconomies respectively.

Why do externalities occur? Roughly, they occur because of the confluence of two features of our condition: (i) our natural world is a tightly integrated (highly interactive or interconnected) — I shall say *coherent* — system; actions taken at any one point reverberate throughout the system, but (ii) socially, our culture is dominated by an economic system in which individuals (persons, companies) compete for economic survival. Economic individuals cannot afford to take into account any of the social costs of their actions other than those for which their competitors can also be forced to be accountable (lest their accounting generosity lead to uncompetiveness and so elimination). Historically, however, we have inherited a culture in which not every, and not even every important, causal repercussion of our typical activities is economically (or even socially) accountable.

It is not too hard to see, at least in outline, how this account extends to the usual list of industral environmental problems. Most obviously, while the air and water remain free dumping grounds for wastes and the earth nearly so, competitors will dump their wastes thus rather than treat and recycle them. Equally, exploitation of the most easily available energy resources now, for purposes which are profitable now, becomes imperative if one is to generate profits and so survive to make future decisions — no matter how inappropriate the usage, how short-term the horizon, how dangerous to civilization the policies: likewise for material resources. Once land, too, is a marketable commodity its use must bend to the momentary dictates of the market, but market purposes are typically in conflict with ecological structure. In similar fashion, the dictates of a continually industrializing agricultural business leads to high-intensity pesticide, drug and fertilizer use. Once again the natural ecology, including both soil ecology and above-surface ecology, are modified (often simplified and destabilized) to suit; the result is, in many instances, plague, deteriorating soil quality, increasing arid areas. Finally, most of the attractive features of urban life are public goods, no one's specific economic concern and so in no one's individual interest to support first (if someone else will do it, why should I?, I can still enjoy the result). Those moving into a city consider only their own advantage in locating, not the overall design of the city; those leaving its centre when the resulting race for downtown space has led to congestion, pollution and high prices do not compensate those remaining, thus leading to a steady exodus and a declining city core (often with the majority recognizing what is happening but each unable *individually* to do anything about it and survive as they desire). Likewise for the urban expansion on to farmland, the increasing spread of expressways and automobile deaths, and so on.

The natural question to ask is "How is it that we generate such external diseconomies so systematically?" What is it about our culture that leads us to institute a set of social arrangements that lead to these kind of difficulties? My tentative answer to these questions requires some considerable preparation. To begin with let me turn to:

> *Proposition 3:* All environmental problems are generated through human situations in which a certain class of games (in the Game Theoretic sense) are realized, of which the game called Prisoner's Dilemma is a paradigm example.

An example of Prisoner's Dilemma is as follows:

	A 1	2
B 1	5,5	1,10
2	10,1	10,10

Each of two "players", A and B, have either of two strategies, 1 or 2, open to them, with the pay-offs for each of the four possible *combinations* of strategies being given in the boxes, in the form A, B. Under the assumption that A and B *independently* choose strategies that *minimize* their individual pay-offs (in the original example the numbers stood for prison sentences), both A and B choose strategy 2. A argues that strategy 2 yields a number as low as, or lower than, does strategy 1, irrespective of what B does; B argues similarly. But the strategy *combination* 2, 2 is clearly the worst of all! *Individual* rational behaviour had led to a *collectively* disasterous result.

The situation we have here, recalling the earlier discussion, can be roughly generalized and described as follows: (i) Each player chooses a strategy which is designed to optimize his own net return from playing the game. But those strategy choices, taken together, lead to a result which is collectively sub-optimal (and often about as bad as one can get in the situation). By acting individually in a rational fashion we bring it about that the collective result is irrational (sub-optimal). (Corollary: in most cases, if all the players were to not act rationally individually, but irrationally agree to cooperate, one can select a set of cooperative strategy choices which leads to each individual receiving more than he could receive by acting rationally and competing.) (ii) The collectively destructive phase of the game sets in when the collective costs imposed on the remaining players by an individual's action exceeds any benefits which all the players together, including the individual in question, gain from the action. (iii) The rules of the game are such that the individual players must ignore the collective costs which their actions impose on others and so must ignore the assessment of the game as collectively destructive whilst ever their own actions bring a net benefit to themselves.

Garrett Hardin's tragedy of the commons is one example of a game of this sort.[8] Though more complex, those in the linear agricultural system play a similar game among themselves, and with us all. And in actual practice the collective costs imposed upon us by the generation of our environmental problems are enormous. To take a particular case more amenable to direct measurement than most, in the late sixties (1968–69) air pollution was estimated to be responsible for $32 billion in property damage to United States citizens and that even a fifty per cent reduction in air pollution would reduce the medical bill by over $2 billion (assuming there to be forty million families in the United States). This figure will now be substantially enlarged. Even a small fraction of that sum invested in air-pollution control equipment would produce a very sizable improvement in the quality of the air, let alone the continuous investment of such sums each year. These games generate high collective costs because the living system of the biosphere is highly coherent, and this web of life is a delicately balanced ecology achieving its present form only after tens of millenia of adjustment, backed by a billion years of evolution. Less easy to measure but of even greater importance are the costs imposed on the human community, in the form of opportunities for choice, flexibility and development lost, by the present pattern of industrial development and the personal costs to people of living under our present urban-industrial conditions.

From this there follows three corollaries:

1. The coherence of our total environment means that an environment- ally compatible (harmonious) society must meet the constraints imposed by the living system (including the human component) if it is to remain healthy; these constraints will be overall (global,

collective) structural constraints on the preservation of energy-matter-information relationships in the system. The achievement of such collective coherence requires an equal degree of social coherence, it requires the ability to achieve a collective social *design* satisfying these constraints. And this in turn demands that human beings assume responsibility for the deliberate design of their social-technological system.

2. The design criteria will all be systems criteria, for the effective functioning of any of our major subsystems (natural-ecological, industrial-technological, socio-cultural) cannot be reduced to just the functioning of their parts; they remain coherent, integrated systems. It is of some importance, therefore, that we support (and even hasten) the development of systems models and design theory in each of these areas.

3. Competitive market relations of the sort which generate the foregoing games are not of themselves capable of achieving collective design coherence of the sort required. And in fact the market generally fragments all decision making into independent, individual short-term decisions and fragments the world into a collection of independent commodities. To impose collective design constraints is to suspend the operation of the market in those respects.

And so the next question to be asked is: How did it come about that we have inherited a society which systematically generates these collectively incoherent games. This leads us to:

Proposition 4: Industrialized market societies have a distinctive institutional design and a correlative culture that characteristically leads to the systematic generation of collective incoherencies (subclass: environmental problems).

To obtain some grasp in short order on the notion of institutional designs, consider the difference between what I shall call a *culturally unitary* society and a *market* society. A culturally unitary society is one dominated by a myth which articulates a conception of the universe and of the nature, significance and destiny of human beings as a group within it, a myth rich enough to articulate a more or less detailed conception of human society and of human persons in their societal context. A market society is one in which the institution of the economic market as the means of allocating resources and distributing goods and services dominates the institutional structure of the society. As a *very* rough approximation we can think of most tribal societies, communal-christian, communist and fascist societies and most utopias as having at their heart a culturally unitary paradigm, while capitalist and moderate-socialist democracies are centrally motivated by a market society paradigm. Real societies are more complex, and less consistent; however, I shall now compare and contrast crude caricatures of these societal alternatives, characterizations crude enough for brevity, but accurate enough for structural insight.

The fundamental unit of the culturally unitary society is the collective whole; its characteristics are paramount, the object of policy. Individual identities and significances are assessed through their relation to the whole, and individual courses of action determined in relation to policy for the whole. By contrast, market society conceives of the individual as the basic unit, autonomous, with internal identity and ends, and sociable only to the degree that this unit finds it self-beneficial to be so. Policy making for the culturally unitary society aims at the realization of the mythically ideal conception of the collective whole and deduces individual policy therefrom; for the market-society collective policy is some weighted sum of individual policies and has no character of its own (it reflects no agreed collective self-image).

For market society, autonomous individuals choose their own life strategies against available knowledge of others' preferences (importantly, estimates of other's preferences in the form of prices). The primary function of government and law is the support of this possibility; hence, government legislates against antimarket activities and the resulting laws are individual, property-ownership oriented with actions based upon market interests. Insofar as collective response is necessary, government decides upon rational action through a mini-market in which individual interests are expressed through various group interests with different weightings than, but in essentially the same way as, they are in the market: this minimarket we call parliament, (prices become votes), and the resulting government, representative democracy. The general character of law remains essentially unchanged, though now with an increased range of property-based collective coercion. Legalized collective coercion of some sort is the basic form of workable solutions to "prisoner's dilemma" games in this context. By contrast, in culturally unitary societies there are no primary individual interests to protect, the main function of government is the formulation of policy which yields and expresses the vision of the collective myth. Consequently, the law, as its vehicle, is primarily concerned with collective rights and collective actions. Government itself may take any decision-making form convenient for this purpose — in particular, because of its collective design defects government is unlikely to take the form of a market.

In the *humanely* efficient market society, education has as its goal the maximization of saleable market skills — educational processes are dominated by institutional forms which hand down information and skills from authority to pupil during the period prior to full-time entry into the market. In the culturally unitary society the aim of education is the creation of the mythically whole man, whatever that demands — it need neither be confined in time, nor to skills and information but may include therapies, cultural practices, etc.

The market economy runs on profits which are functions of input-output differences computed at the individual level for populations of commodities. Economics for the culturally unitary society is based on the regulation of the collective access to resources so that the collective productivity may realize the ideal states of the collective myth. This latter approach can lead to many different political economies depending, among other factors, on the myth involved; but often it will focus around a process-stock economy where the fundamental economic units are structures of related living processes that are to be preserved and the fundamental constraint is the future-oriented conservatory management of available stocks. In fact, in a culturally unitary society the activity we would describe as economics is not conceived of as a separate group of economic exchanges; that is, a distinct area of social activity governed by its own laws, not subject to interpretation in terms of social obligation — *caveat emptor.*

It is clear that to participate happily and wholeheartedly in the market society requires the willingness to adopt a particular view about the nature of human beings and the goals of their lives; that is, to adopt a particular set of values and ethical principles. There will certainly be support for some traditional market values: respect for law, honesty, responsibility, reasonableness and so on, for only through the exercise of these virtues can the market function most effec.ively. But there are other, deeper-going value-judgements involved as well. For example, one must be willing to bend the individual's place of residence, living environment, job, and so on to the dictates of the market. At a somewhat deeper level, one must be willing to agree that the essential nature of justice is to protect the rights of individuals in the competitive market (rather than, say, the preservation of minimal living standards for everyone, or the achievement of collective communal qualities above individual interests), agree that it is more important to maximize an individual's marketable abilities than to concentrate upon his refinement in the image of a preferred myth and agree that the essence of human freedom is the removal of social restraint on individual action rather than the societal bestowal of uninherited characteristics. Still deeper, one must be willing to agree that the conception of ourselves as autonomous individuals whose primary social relations are expressed through market activity is the most satisfying human self-image.

Because its self-conception is based on the notion of autonomous individuals, there is a strong tendency to believe that market society itself is free of collective values and judgments. Indeed, one of the chief platforms of its defense over the last two centuries was precisely that this social form freed people from coercive dogma: religious, ethical and social. There is an important element of truth to this. However, the foregoing brief consideration of market society should indicate that it, too, is based on a pervasive myth, though in a subtly different way. The myth in question involves a conception of human nature, our society and the destiny of both. Within a market society, the market mechanism is not a function of the value judgments of *individuals,* but the choice *of* a market society as the preferred social form is an ultimate evaluative social choice.

And so one could continue. The value of the analysis is not to uphold some preferred society, for even approximately culturally unitary societies have historically either had deep defects (e.g., been repressively totalitarian) or are separate from us by an educational-technological gulf. The value of the analysis is the way in which it permits us to escape the conceptual confines of our culture and enquire into alternative social designs. Beyond the simple-minded market/totalitarian dichotomy lies the serious field of psycho-social institutional design. (Only our present culture and the fact that our conceptions of design in the latter areas lag behind our natural science by thousands of years block our perceiving this.)

The characteristic institution in market society is the private enterprise and, generalizing, since even public institutions tend to be modelled after them, I shall call them "commodity-interventions" (CI) institutions because each is so designed as to efficiently regulate the numbers in a single population of commodities. If we model institutions as graphs with the nodes as decision/information-transformation matrices representing persons and the lines joining them as lines of communication and decision flow, then CI institutions are unidirectional convergent nets (they look like pyramids and the information flows only up, with variables integrated out — filtered — at each stage and decisions flow only down, dually fragmented through conditional-

ization on the integrated variables at each stage). Such institutions lead
to specialized human roles; the institutions show little intelligence in
any area except for the population on which they are focused and no
relevant external adaptability (though they often do grow and survive,
even when it is not in the collective interest). Because, in a market
society, governments aim not to intervene in the market and because
all individuals have absorbed the market paradigm, public institutions
tend also to be focused on the control of commodities and to be
designed like CI institutions. Examples are the authority communica-
tions structures of private enterprises, hospitals, electricity authorities.
Market institutions effectively fragment their world into its individual
commodities.

The resulting fragmentation of conception and decision making is
worth a more explicit remark. The market society dictates decisions
made at the individual level, primarily in the short-term perspective,
with the only relevant information being (systematically) partial information concerning those priced commodities (i.e., objects, not
processes) immediately involved. This circumstance involves a three-
fold fragmentation of our decision making. (i) The entire web of life,
human and natural, is not treated as a structure of dynamic processes
but as a collection of separate objects — commodities (trees, beaver,
wheat, oil, etc.) — which are tradeable and therefore are given market
prices. (ii) Entire slabs of the ecology are eliminated from consideration
altogether because they go unpriced (air, water, wildlife). (iii) The
historical time scale often crucial to grasping the dynamics of living
processes is fragmented into a myriad of short-term segments. Turning
to decision making: (i) Ecologically systematic and or collective
decisions are fragmented into myriads of individual decisions. (ii) Only
individual outcomes, not collective outcomes, are taken directly into
account. These fragmentations crucially distort both the environment
and rational decision making. These fragmentations form the basis for
the ubiquitous emergence of "prisoner's dilemma" games, market
competitiveness undergirds the choice of individually optimal strategies
evaluated on these fragmentations. What one expects to find are linear
systems designed (as in the agricultural example discussed earlier)
with individual CI institutions existing on particular input/output
differences for individual segments in the linear chain. And all this leads
to:—

> *Proposition 5:* CI institutions lead to systematically misposed
> questions (problems) and so to systematically misleading
> responses.

In market society it is unavoidable that problems will be posed on the assumption of fragmentation where there is none, and strategies devised on the assumption that fragmented decision making can yield optimal results, whereas it cannot. Examples have already been provided. Another example is: how to raise the efficiency of emergency health care services for motor-vehicle accidents as the *primary* social response to road carnage (when it is the design of the motor-vehicle system, engineering and human, which is largely responsible).[9]

In light of the foregoing, the natural concept around which to organize theory and practice is that of *design*. Ecologies, industries, institutions and personalities are all complex, coherent systems – it is their design that is their key feature. Good environmental policy is policy consciously designed to effect a complex design in society or maintain and/or enhance an ecological design in nature. Consciousness of misposed issues brings consciousness of poor design: houses built in ignorance of climate, the high-rise office tower's heating/cooling and transportation inefficiencies, and so on. The more complex the system the greater the role of information flow in determining function, stability and other key features. Among the many distinctive features of this century the one that stands out in my mind is the conscious development of theories peculiarly concerned with design. Hitherto humankind has simply exploited energy systems, first as adapted predator, then as adapting sedentary agriculturist and finally as industrialist using external (to photosynthetic) energy subsidies to re-create the environment as a human artifact. Now at least we have taken the next significant step and begun the science of design without which the artifactual world is blind. Environmental problems are vivid foci of these developments.

There *are* hints of the design theme in Passmore's writing. He hints at some of the design complexities really buried in a population policy – not just simple numbers of people are involved, but patterns of land use and urban design are at stake, the design of social freedoms, and so on (p. 155). And by page 183 – but that is near the end – we hear for the first time about a (Heracliton) Western tradition in which "the world consists of complex systems of interacting processes varying in their stability", which is in contrast to "that Aristotelian-type metaphysics for which nature consists of substances with distinct properties". And on page 178 we read: "Men, uniquely, are capable of transforming the world into a civilized state, that is their major responsibility to their fellow man." By page 188 we read that "certainly there is little hope for us unless we can modify our desire to possess. We shall do so, however, only if we can learn to be more sensuous in our attitude to the world". Here we have in these quotes, taken successively, tentative remarks in the direction of a theory of metaphysics, a theory of the significance of humankind and of the nature of the planet as a human artifact, and a theory of human psychology and culture. These together might form the elements of a coherent environmental world view. The theoretical level of treatment is light; it is indicated by the quote at the outset of this paper. Nonetheless, each of these elements distinctively involves design; that is the key concept in the understanding of each element. What a pity we hear about them only in snatches, at the book's close. (Ironically, these remarks, among the more innovative in the book, occur in a last chapter entitled "Clearing away the Rubbish", suggesting a mere clarifying role for philosophy once again.)

I do not come to this essay with a prepared utopian design to offer — I do not believe such exists. I can contribute only the following:

Firstly, an historical awareness of the urgency and fundamentalness of the design issue. This I have attempted to underscore throughout the preceding.

Secondly, some general suggestions concerning approach. Consider first Commoner's position.[10] As a first approximation, we can view our activities in three interconnected systems — ecology, technology and economy/socio-polity. At present the priority relations run from the last of these to the first, while the real dependency relations are from first to last. (E.g., technology is really dependent upon ecology to sustain itself, but we give technological advantages priority over ecological consequences in our decision making.) In consequence we have a systematically misdesigned technology which does not serve human ends well, and disrupted ecological systems. The design priorities must, then, be reversed; the technological subsystem needs an ecologically compatible design and the economy/socio-polity a design which will reinforce the former design imperative and, simultaneously, promote human value and well-being.

This has led some friends and I to focus on what we call *match* (M) institutions in which, in contrast to CI institutions, the objective is to manage a complex system (e.g., some aspect of the ecology-techno-economy), maintaining system integrity; the general means chosen is to reflect the system structure in the institutional structure itself. Politically we also look towards what we call criterial (C) institutions, which extend the parliamentary process, not towards the traditional civil service, but towards public and judiciary (without comprising judicial processes) after the general manner of boards and royal commissions (cf. Passmore's citing of a National Resources Council proposal, p. 98). And socio-politically we emphasize interdependent decentralized institutions as a counterweight to the natural concentration of power resulting from increasingly large scale and complex decisions. Most of this position remains to be worked out, and much of it can in fact only be worked out through experiment. Both theoretically and experiencially we humans are as yet only "babes in the woods" — and yet there are many interesting fragments of theory and practice to draw upon.[11]

Thirdly, from an abstractly theoretical point of view I can point to what I think of as the true excitement of the twentieth century and to some exciting theoretical prospects beyond the more immediate institutional concerns at which theory was aimed in the previous paragraph. The twentieth century is marked by new theories in physics, certainly, but in my view its really striking feature is the emergence of major theoretical areas devoted wholly to the design and control of complex systems: operations research, decision theory, general systems theory, network analysis, control theory, cybernetics, artificial intelligence, and so on. Much could be written on these developments and their implications, especially the fruitful interdisciplinary relations now emerging. Here I mention only some foci: in self-organizing systems (biology, ecology, artificial intelligence), resiliency and surprise (ecology, decision theory), distributed control (operations research, artificial intelligence), heirarchy and control plasticity (cognitive neuropsychology, biology, engineering).[12] None of this theory is yet near to offering interesting specific proposals for institutional design, but we may hope that some day our societal self-control will begin to approach the elegance, integration and local autonomy of, say, the human central nervous system.

Some day, soon, I believe, systematic design must be taken very seriously if humaneness is to survive knowledge.

Notes

1. For a discussion of this metaphilosophical framework see C. A. Hooker, "Systematic Realism", *Synthese* 26 (1974):409–97, and idem, "Systematic Philosophy and Meta-Philosophy of Science", *Synthese* 32 (1975):177–231.
2. Aspects of my own environmental/public-policy approach can be found in A. Hanna and C. A. Hooker, "Towards a Re-conception of Public Policy" to appear in Festschrift for Sir Geoffrey Vickers; C. A. Hooker et al., *Energy and the Quality of Life: Understanding Energy Policy in Canada* (Toronto: University of Toronto Press, 1980); C. A. Hooker, "Cultural Form, Social Institution, Physical System: Remarks Towards a Systematic Theory of Environmental Problems", in *Proceedings of the Second International Banff Conference on Man and His Environment,* ed. M. F. Mohtadi (New York: Permagon, 1975); C. A. Hooker and R. van Hulst, *Institutions, Counter-Institutions and the Conceptual Framework of Energy Policy Making in Ontario* (Research Report to the Royal Commission on Electric Power Planning, Ontario, Canada, May 1977); C. A. Hooker and R. van Hulst "The Meaning of Environmental Problems for Public Political Institutions", in *Ecology Versus Politics in Canada,* ed. William Leiss (Toronto: University of Toronto Press, 1979), pp. 130–63; and C. A. Hooker and R. van Hulst "Institutionalising a High Quality Conserver Society", *Alternatives* (Journal of Friends of the Earth, Canada) 9 (Winter 1980). Related to science and epistemology, see C. A. Hooker and A. K. Bjerring, "The Implications of Philosophy of Science for Science Policy", in *The Human Context for Science and Technology,* vol. 1, ed. C. A. Hooker and E. Schrecker (Ottawa: The Social Sciences and Humanities Research Council, 1980); C. A. Hooker, "Formalist Rationality: The Limitations of Popper's Theory of Reason", in *Metaphilosophy,* 12 (1981) 248–66; C. A. Hooker, "Scientific Neutrality Versus Normative Learning: The Theoretician's and the Politician's Dilemma", *Science and Ethics,* ed. D. Oldroyd (Sydney. University of New South Wales Press, 1982).
3. Even so, the conservatism is there. The final policy Passmore adopts, namely, handing on the planet to the next generation in the best possible condition, neglects many factors, for example, cyclic phenomena in nature whose periods are longer than one generation – unless "best possible condition" is given a very wide reading indeed (but see, e.g., p. 121 for an implied commitment to a wider criterion). But then a much deeper and, I shall now argue, more radical analysis is needed to understand the criteria.
4. The cries against aesthetic pollution are widespread. For antichauvinist ethics see Richard and Val Routley, "Human Chauvinism and Environmental Ethics", in *Environmental Philosophy,* ed. Don Mannison, Michael McRobbie and Richard Routley (Canberra: Research School of Social Sciences, Australian National University, 1980); for the Jain religion see Ninian Smart, *The Religious Experience of Mankind* (London: Collins, 1968), Chardin's somewhat controversial Catholic-evolutionist views are expressed in P. T. de Chardin, *The Phenomenon of Man* (London: Collins, 1961) among other books; and Lewis's stauchly Protestant Christian views are vividly conveyed in C. S. Lewis, *That Hideous Strength* (London: Pan, 1955); and idem, *Poems* (London: Bles, 1964).
5. I say "would have" because, although this is probably the course that will be chosen by the country's decision makers, the situation is currently (1982) being re-evaluated, spurred by the growing demand for a long-term, nonincremental evaluation and by the current recession-induced decline in the rate of increase of electrical demand.

6. The idea of retrospective path analysis was developed from the approach of Lovins to energy policy, see, for example, Amory Lovins, "Exploring Energy-Efficient Futures for Canada", *Conserver Society Notes* 1, no. 4 (1976): 5–16; and idem, *Soft Energy Paths* (Cambridge, Mass.: Friends of the Earth, Ballinger, 1977). However, Lovins takes a moderately narrow technical approach and I, and a former student of mine, Hanna, attempted to generalize it to a broadly applicable policy. See A. Hanna, *Settlement and Energy Policy in Perspective: A Policy Evaluation* (Ph.D. Thesis, University of Western Ontario, Canada, 1980); and Hanna and Hooker, "Re-conception of Public Policy", for a summary. It has been applied to alternative or "soft" Canadian energy policy recently in the coordinated studies presented in *Alternatives* (Journal of the Canadian Society of Friends of the Earth), 8 (Summer/Fall 1979) and 9 (Spring 1980). For recent discussions of cost-benefit criteria see, for example, Stafford Beer, *Platform for Change* (New York: Wiley, 1975); C. Nash et al., "An Evaluation of Cost-Benefit Analysis Criteria", *Scottish Journal of Political Economy* 22 (1975):121–24, Mancur Olson, "Cost-Benefit Analysis, Statistical Decision Theory, Environmental Policy" in *PSA 1976*, vol 2, ed. Patrick Suppes and P. D. Asquith (East Lansing, Michigan: Philosophy of Science Association, 1977); C. Starr, "Social Benefit Versus Technological Risk", *Science* 165 (1969): 1232; and especially the article by Amory Lovins, "Cost-Risk Benefit Assessments in Energy Policy", *George Washington Law Review* 45 (1977):911–43.

7. For more on this theme see, Hooker et al., *Energy and the Quality of Life*, chaps. 10, 11, and 34–36.

8. See Garrett Hardin, "The Tragedy of the Commons", *Science* 162 (1968): 1243–48. The argument is explicit in many books on the economics of environmental problems, for example, P. W. Barkley and D. W. Seckler, *Economic Growth and Environmental Decay* (New York: Harcourt, Brace and Jovanovitch, 1972); and implicit in many more, for example, H. E. Daly, *Steady-State Economics: The Economics of Bio-Physical Equilibrium and Moral Growth* (San Francisco: W. H. Freeman, 1977).

9. The literature in the general area of environmental problems is vast and of uneven quality. In energy policy alone I and my friends have listed nearly 500 relevant works in Hooker et al., *Energy and the Quality of Life* – and this bibliography was concentrated specifically on Canadian energy policy. Given my own general analysis of environmental problems, I recommend (though usually not unreservedly), Barkley and Seckler, *Economic Growth*; Wandell Berry, *The Unsettling of America: Culture and Agriculture* (San Francisco: Sierra Club, 1977); H. D. S. Cole et al., *Thinking about the Future, a Critique of the Limits to Growth* (London: Collin, Chatto and Windus, 1973); David Collard, *Altruism and Economy: A Study in Non-Selfish Economics* (New York: Oxford University Press, 1978); Barry Commoner, *The Closing Circle* (New York: A. A. Knopf, 1971); idem, *The Poverty of Power: Energy and the Economic Crisis* (New York: Bantam, 1977); Daly, *Steady-State Economics*; W. Deane, ed., *Growth in a Conserving Society*, (Toronto: York Minster, 1979); David Dickson, *Alternative Technology and the Politics of Technological Change* (Glasgow: Fontana, Collins, 1974); Mark Diesendorf, ed., *Energy and People: Social Implications of Different Energy Futures*, (Canberra: Society for Social Responsibility in Science, 1979), Steven Ebbin and Raphael Kasper, *Citizen Groups and the Nuclear Power Controversy: Uses of Scientific and Technological Information* (Cambridge, Mass.: MIT Press, 1974); Samuel S. Epstein, *The Politics of Cancer* (San Francisco: Sierra Club, 1978), Victor Ferkiss, *Technological Man* (New York: Mentor, 1971); Nicholas Georgescu-

Roegen, *The Entropy Law and the Economic Process* (Cambridge, Mass.: Harvard University Press, 1974); Hanna, *Settlement and Energy Policy*; Hanna and Hooker, "Re-conception of Public Policy"; H. Henerson, *Creating Alternative Futures: the End of Economics* (New York: Berkley, Wyndhover, 1978); Hooker, *Energy and the Quality of life*; Hooker and van Hulst, *Institutions*; idem, "Environmental Problems"; idem, "High Quality Conserver Society"; Ivan Illich, *Tools for Conviviality* (New York: Harper and Row, 1973); William Leiss, *The Domination of Nature* (New York: Braziller, 1972); idem, *The Limits to Satisfaction* (Toronto: University of Toronto Press, 1976); Lovins, "Energy-Efficient Futures"; idem, *Soft Energy Paths*; idem, "Cost-Risk Benefit"; Donella H. Meadows et al., *The Limits to Growth* (New York: Universe Books, 1972); Hugh Nash, ed., *Progress as If Survival Mattered* (San Francisco: Friends of the Earth, 1978); William Ophuls, *Ecology and the Politics of Scarcity* (San Francisco: W. H. Freeman, 1977); D. C. Piraeges, ed., *The Sustainable Society: Implications for Limits to Growth* (New York: Praeger, 1977); Karl Polanyi, *The Great Transformation* (Boston: Beacon Press, 1960); J. Robertson, *The Sane Alternative: Signposts to a Self-Fulfilling Future* (London: James Robertson, 1978); Routley and Routley, "Human Chauvinism"; E. F. Schumacher, *Small is Beatiful: A Study of Economics as if People Mattered* (London: Abacus, 1974); E. F. Schumacher and P. N. Gillingham, *Good Work* (New York: Harper and Row, 1979); Bruce Stokes, *Helping Ourselves: Local Responses to Global Problems* (Washington D.C.: World Watch Institute, 1978); A. J. Surrey et al., *Democratic Decision-Making for Energy and the Environment* (London: Macmillan, 1978); Robert Theobald, *Habit and Habitat* (Englewood Cliffs, N.J.: Prentice-Ball, 1972); idem, *Futures Conditional* (New York: Bobbs-Merrill, 1972); Lawrence Tribe, "Technology Assessment and the Fourth Discontinuity: The Limits of Instrumental Rationality", *Southern California Law Review* 46 (1973):617–60; Geoffrey Vickers, *Freedom in a Rocking Boat* (London: Penguin, 1970); D. White et al., *Seeds for Change, Creatively Confronting the Energy Crisis* (Melbourne: Patchwork Press, 1978); M. C. Yovits et al., *Self-Organizing Systems* (Washington: Spartan Books, 1962). I have already referred to my own writings in n. 2, above.

10. Commoner, *Poverty of Power*; see also his earlier *Closing Circle.*

11. Practice ranges from experiments in local worker control in Yugoslavia (see, e.g., G. Hunnius, "The Jugoslav System of Decentralization and Self-Management", in *The Case for Participatory Democracy*, ed. G. Bennello and G. Roussopoulos [New York: Viking Press, 1971]) to the emergence of environmental impact statements as a major planning tool in New South Wales, Australia. Does any institution, anywhere, have the foresight and structure to begin studying these ranges in the whole? For my own views see n. 7, above.

12. For an introduction to this theoretical literature see, W. R. Ashby, *An Introduction to Cybernetics* (London: Chapman and Hall, 1968); Stafford Beer, *Platform for Change* (New York: Wiley, 1975); idem, *The Heart of the Enterprise* (New York: Wiley, 1979); C.W. Churchman, *The Design of Enquiring Systems* (New York: Basic Books, 1972); F. E. Emery and E. L. Trist, *Towards a Social Ecology* (New York: Plenum, 1973); Nicholas Georgescu-Roegen, *The Entropy Law, and the Economic Process* (Cambridge, Mass.: Harvard University Press, 1974); C. S. Holling, ed., *Adaptive Environmental Assessment and Management* (New York: Wiley, International Institute for Applied Systems and Analysis, 1978); Hooker, "Cultural Forms"; Erich Jantsch, *Design for Evolution* (New York: George Braziller, 1975); idem, *The Self-Organizing Universe* (New York: Pergamon, 1980); Erich Jantsch and C. H.

Waddington, *Evolution and Consciousness: Human Systems in Transition* (London: Addison-Wesley, 1976); Humberto R. Muturana and Francisco Varela, *Autopoiesis and Cognition: The Realization of the Living* (Dordrecht: Reidel, 1980); Meadows, *Limits to Growth*; I. I. Mitroff, "Systems, Enquiry and the Meaning of Falsification", *Philosophy of Science* 40 (1973):255–76; G. Nicolis and I. Prigogine, *Self-Organization and Non-Equilibrium Systems* (New York: Wiley-Interscience, 1971); Howard H. Pattee, *Hierarchy Theory* (New York: Braziller, 1973); J. A. Simon, *The Sciences of the Artificial* (Cambridge, Mass.: MIT Press, 1969); Francisco Varela, *Principles of Biological Autonomy* (New York: Elsevier, 1979); Yovits et al., *Self-Organizing Systems.*

Preservation of Wilderness and the Good Life

*JANNA L. THOMPSON**

I

> The day shall dawn when never child but may
> Go forth upon the sward secure to play.
> No cruel wolves shall trespass in their nooks,
> Their lore of lions shall come from picture books.
> No ageing tree a falling branch shall shed
> To strike an unsuspecting infants head.
> From forests shall be tidy copses born
> And every desert shall become a lawn.

Christmas in New Rome (1862)

What is objectionable about this vision of a human-made Garden of Eden? We do know that grandiose schemes for restructuring the earth can have unforeseen and harmful consequences for human life and health. But human survival would probably not be affected adversely by the extinction of lions and wolves. Large areas of wilderness could no doubt be converted to farmland, pasture, mines, cities or parks without endangering our species. Potentially useful plants or animals might be kept in botanical gardens, laboratories or zoos. We might even find ways of sustaining ourselves in comfort without having to preserve wilderness at all. Let us assume that such "optimism" is justified. Are there any grounds left for a protest against the destruction of wilderness, species and natural ecosystems?

If wilderness is not needed for our physical salvation, then perhaps it

* I would like to thank those people who have given me help and encouragement in writing this paper, particularly H. J. McCloskey and John Campbell of La Trobe University Philosophy Department.

is needed as salve for the spirit. Almost every campaign for wilderness preservation makes an appeal, sooner or later, to the psychological value of wilderness:

> The thought of "the calm, the leaf and the voice of the forest" is itself a refuge from stress, a wilderness at the back of the strained mind. When we finally know that the last forest has gone, that there is nowhere to go but along the runways of our steel and concrete anthills, that the last link with our past has snapped, then perhaps we may snap too. We will have no refuge left at all.[1]

In *Man's Responsibility for Nature,* John Passmore denies that this need for wilderness is a general need.[2] Some people do feel diminished by the destruction of wilderness and species, but others do not. Many people, including Passmore himself, would prefer a park to an impenetrable forest, a lawn to a desert. In any case the need for wilderness has not always existed, and in the future it may no longer exist. The idea that wilderness has a psychological value was almost unheard of before the romantics of the eighteenth and nineteenth centuries made the idea fashionable. But fashions come and go. Future generations may not value wilderness – a consideration that diminishes further, in Passmore's eyes, the strength of the case for preservation.[3] If we're not sure whether future people will have a particular need, then surely our obligation to provide for it diminishes.

If the psychological need for wilderness comes down to the desire of a few people living at a particular time for the kind of experiences which a wilderness can provide, then the case for the preservation of wilderness on this ground does indeed look rather weak. The interests of a few bushwalkers and naturalists seem rather insignificant when compared with the material interests of people in the forestry, grazing and mining industries who can say, with some justification, that their activities benefit large numbers of people; or when compared with recreational needs of people who want roads and golf courses rather than wilderness.

From another direction, Richard and Val Routley also raise doubts about the effectiveness of the psychological case for the preservation of wilderness.[4] What is actually being valued, they suggest, is not wilderness, but the mental states it produces: feelings of awe, experiences of beauty. But if the need for wilderness is simply the desire to have such experiences, then it seems possible that whatever people get out of wilderness can be provided by some substitute. Films of wilderness might be sufficient or, if not, someone might invent the "wilderness-experience machine" — a device which, when attached to the head, gives you all the sensations of being in a wilderness in your own living room. Actual wilderness need not exist at all.[5]

Making a good case for the preservation of wilderness seems to require a departure from standard ways of dealing with ethical issues. Some preservationists, like the Routleys themselves, contend that what is needed is to break away from those systems of normative ethics, according to which human individuals and their states — or by extension, sentient beings and their states — are the only things of ultimate value.[6] Others argue that we must question in a fundamental way, commonly held views about what human welfare and happiness consists of.[7] But the pursuit of a new environmental ethic runs up against what appears to be an insurmountable obstacle. The problem faced by those who advocate a radical departure from traditional normative ethics is encapsulated in the dilemma described below.

People value what they value. But to deserve the name of an ethic, a position on values must be persuasive. Either a new normative ethic of the environment will rest its appeal on the principles, values or ideas about their own interests that people already accept, or it will not. If it does not depart from these accepted views then it will fail to provide an adequate case for the preservation of wilderness and species. But if it does depart from traditional axioms, ways of arguing or ideas about interests, then it won't be persuasive. So traditional appeals and ways of arguing won't do and new ways seem to be ruled out.[8]

II

How can a revolution in ethics be possible? Radical changes seem to be ruled out by a tacitly accepted requirement that ethical positions not depart substantially from given ethical premises and ways of arguing. But why should this requirement be imposed on ethical innovation? Why should radical change, which we acknowledge to be possible elsewhere, be excluded from ethics? The answer seems to lie not in the nature of ethical discourse, but in a widely accepted approach to ethical issues.

The debate about environmental ethics has taken place almost entirely within the isolated confines of the area of philosophy called "ethics". Inside this framework appeals are made to recognized ethical principles, categories and metatheories; terms like "intrinsic value", "instrumental value", "good", etc., are freely used; certain vague but presumed self-evident values like happiness, pleasure, self-fulfillment are referred to, and people's ethical perceptions and their ideas about their interests are taken as given.

This way of treating ethical matters has some unfortunate consequences. The first is that it segregates ethical issues from other considerations: from scientific knowledge, for instance, or from analyses of political change and social life. This segregation is sometimes justified as a consequence of the separation between "is" and "ought". But the defense of this distinction does not require us to deny that ethical axioms can and should be influenced by new knowledge about the world or by attempts of people to understand and deal with discomforts and conflicts in their lives.

Another related consequence of a compartmentalized ethics is that principles and categories tend to be presented ahistorically as if they were unchanging and unchangeable. The only alteration that is really allowed for is a change in ethical sensibilities: a greater willingness to accept the consequences of ethical principles that already exist.[9]

Passmore, in *Man's Responsibility for Nature,* does undertake a historical survey of Western attitudes towards nature and is thus aware that substantial ethical changes can occur. He believes, for instance, that changes in ideas about our obligations to animals are the result of a real ethical change.[10] On the other hand, he treats developments in the history of ethics as if they were only marginally connected with other historical developments, and this limits the changes he can allow for as far as environmental ethics is concerned. Only evolutionary change seems possible — the changes which allow the new to be seen as a slight variation of the old. A radical break with tradition is bound to seem impossible as long as we confine ourselves to a view of ethics which gives us no way of comprehending such a break.

Passmore's way of dismissing the notion that wilderness has psychological value illustrates the limitations of his approach. For him the idea that wilderness has this kind of value is something that materializes with the romantic movement. He does not regard it as part of his job as a philosopher to examine the social critique behind the romantic movement. But, detached from its social origins, the idea that wilderness has psychological value can only be treated as if it were a new fashion, which might just as easily go out of season in a generation or so. The need for wilderness thus belongs only to the few people who actually express it, and cannot be seen as the romantics intended — as

a general need, acknowledged or not, which arises from the dissatis-
factions and conflicts of an industrial society.[11]

In the pursuit of an environmental ethic we need, therefore, to
venture outside the area designated as "ethics"; we have to take into
account world views, empirical hypotheses and social analyses. Unfortunately, even advocates of a new ethic do not generally do this. In
"Human Chauvinism and Environmental Ethics", the Routleys argue
that changed attitudes towards the environment make a new environmental ethic possible.[12] Their critics tend to ignore or dismiss these
changes. But because of the self-imposed limitations of ethical
discourse, neither they nor their critics ask the crucial question,
namely, where do these new ethical attitudes come from? The debate
about the possibility of a new ethic thus ends in a stalemate. The critics
possess the tactical advantage of having the weight of our ethical
tradition behind them. But no real progress can be made by assuming
that the conservative position has to be the right one.

III

The change of ethical consciousness, which the advocates of a new
environmental ethic make so much of, goes beyond the awareness that
you have to be nice to the environment so that it will continue to be
nice for you. The nature of the change is shown in the answers that
people of altered consciousness would give to the "hypothetical situations" which the Routleys have devised.[13]

People who really do value wilderness and species, they suggest, will
not find the wilderness-experience machine, or anything else, an
adequate replacement for wilderness. People who really value wilderness and natural systems will not think it morally permissible for the
last people on earth (who know they are the last people) to set about
destroying the plant and animal species of their world.[14] People with
the new attitudes will avoid behaviour which seems perfectly all right
to people who don't have these attitudes. For example, they "may
avoid making unnecessary noise in the forest, out of respect for the
forest and its nonhuman inhabitants".[15] People with the new values
will disapprove of certain ways of using natural systems and living
creatures. They are the people who would continue to criticize factory
farming even if it were discovered that some kinds of chickens don't
mind a battery existence.[16]

According to the traditional human-centred ethic, which Passmore defends, these ethical attitudes don't make much sense. If wilderness is valuable because of the feelings it inspires, then why shouldn't a wilderness machine be an adequate substitute? If the value of plants and animals is to be useful to humankind, then how can we object to what the last people do? Even an ethic broadened to incorporate the welfare of other sentient creatures cannot accommodate the new evaluations.

The evaluations elicited by the Routleys' hypothetical situations could be regarded as test cases for any proposed environmental ethic. But like many other advocates of a new ethic, the Routleys believe that the only one which will actually pass these tests is an ethic which takes certain properties of environmental systems as a whole to be "intrinsically valuable": "Diversity of systems and creatures, naturalness, integrity of systems, stability of systems, harmony of systems."[17]

As intrinsic values go, the Routleys' "multiple factor value" is reasonably well defined. One of the accomplishments of ecology as a science has been to provide an understanding of what terms like "diversity" and "stability" mean when applied to environmental systems. But by making these general properties of natural systems into an intrinsic value, the Routleys construct an ethic which is *too* far removed from human attitudes, feelings and ways of life. The very remoteness of this value from human concerns means that appeals to the diversity, integrity and stability of systems are going to be inadequate to condemn those human activities which many people want to condemn — the destruction or misuse of the wilderness and species that we share the earth with.

Like others who advocate a new environmental ethic, the Routleys do not think that we should aim to preserve whatever now exists. But a change is bad in their view if it brings about a state of the environment that is less valuable than the preceeding state — that is, if the new ecosystems which result from the change are less stable, diverse, have less integrity than the ecosystems which originally existed. After a change, there will, of course, be a period of time of longer or shorter duration when the environment will be in a degraded state. But it is not clear why a change should be judged by its immediate results. If the end result is a state of affairs which has equal or greater value on their criterion of value, then how can the change which produced it be regarded as a bad thing?

In the late Cretaceous, seventy million years ago, the earth was devastated by one of the worst cataclysms in the history of the planet. No large land animal was left in existence; an enormous number of species of plants and animals were destroyed in the sea as well as on land. The forms of life remaining were enormously depleted in numbers, and they were left on an earth that would have seemed empty.[18]

Let us suppose that this destruction was caused by intelligent beings from another planet who used Earth as a testing ground for neutron bombs. Were their actions wrong? The immediate effect was, of course, severe degradation of ecosystems. But after several million years, new systems of plants and animals established themselves. It would be hard to argue that in the very long term, the postcatastrophe environments were any less valuable than the precatastrophe environments. For similar reasons we might argue that the last people would not be wrong to destroy large parts of their planet, providing they did not destroy everything. But what preservationists want to condemn is our lack of concern for the environmental systems and species that now exist on earth. They would not condemn human actions less if they believed that after a million years or so, the earth could recover from human devastation. What the preservationist really wants is for us to come to value a more respectful, harmonious way of living with nature. But this result by no means follows from an insistence that some properties of natural systems are intrinsically valuable.

Nor does the acceptance of the Routleys' position, even among people who take it seriously, obligate us to behave in a different way towards the environment. Logically speaking, it does not follow that we would ever have to prefer environmental values over other things that we hold to be intrinsically valuable — like human happiness and welfare. In practice, we know that even those who profess a love of nature often show by their actions that they in fact value their own welfare and happiness a lot more. An environmental ethic often seems to be something we can show off on Sundays and holidays; the rest of the time we engage in those activities which presuppose the continued destruction of wilderness and species, and we fall into the ways of thinking and attitudes appropriate to these activities. This is not a matter irrelevant to ethics. A way of life may not only make it difficult to practice an ethic; it may also encourage attitudes which ensure that any proposal for a new ethic appears incomprehensible. Therefore, a new environmental ethic cannot simply amount to a proporal of a new system of values. A revolution in ethics also requires a revolution in ideas about social existence.

The new environmental consciousness does, in fact, often go along with a desire for a new way of life. The Routleys themselves demonstrate this in a later work, "Social Theories, Self Management and Environmental Problems". They argue here that people in a society with a radically different economic and political system — an anarchist society — will not only be able to live more satisfying lives but will be able to develop "a genuinely environmental ethic and ecological outlook".[19] Human welfare no longer opposes itself to values in the environment. The way people live is supposed to encourage respect for natural systems. Reciprocally, respect for natural systems enhances personal development:

> Just as the social individual's interests and concerns and welfare are thus bound up with those of others, and are not contained in those of the narrow self of egoism, so the environmental individual's interests, concerns and welfare are bound up with those of the natural world or aspects of it. The result is, as in the social case, a broader concept of self ("I am the eagle") in which the self is connected to the natural, non-human world as well as to other humans, and is able to reach out beyond itself via appropriate relations.[20]

Ideas about new social relations bring with them new conceptions about what human welfare and happiness consist of. It is these new conceptions which make a new environmental ethic possible. However, it is by no means clear that the environmental ethic which is made possible by these changes needs to be, or can be, the intrinsic-value ethic which the Routleys have elsewhere promoted. Indeed the usual distinction between intrinsic and instrumental value, on which proposals of environmental ethics so commonly hinge, seems to be inadequate to deal with the kind of ethical change that is taking place.

In discussions about intrinsic values, it is generally assumed that things of intrinsic value are fixed points to which instrumental values, and other intrinsic values, are contingently related. Instrumental values are tools which generally, though not inevitably, achieve some good that is self-evident. This conception is only possible when these values and their relationships are frozen in the timelessness of philosophical discourse. In practice, intrinsic values prove subject to change and redefinition. They can be reshaped to incorporate what was hitherto regarded as an instrumental value. They can be adjusted to be compatible with each other or to be the sort of things which particular instruments can achieve.[21]

This is what happens in the Routleys case. They deliberately "construct" a society so that human happiness — one intrinsic value — can become compatible with, and even conducive to, the promotion of environmental values. Similarly, though less noticeably, their environmental values are given a human shape. The environment we are supposed to come to value is our environment: the land around us and the species with which we share the land. But why should *this* environment and *these* species be especially valuable? The intrinsic value ethic gives no answer to this. But the obvious reply is that these particular natural systems should be respected and preserved because of their special relationship to us, and to our mental and physical well-being.

Since the new environmental consciousness is not adequately catered for by an ethic which detaches itself completely from human needs and desires, I suggest that we look for an environmental ethic which brings together two approaches that are usually thought to be incompatible. First of all, this new environmental ethic will revive and incorporate what I have called the psychological case for preservation — the idea that human self-realization and happiness depend on developing a new relationship with the environment. Secondly, it will incorporate what I think is correct in the Routleys' position: that there is something wrong with a view which sees the natural environment and living creatures as being valuable only as instruments for promoting human well-being. The view I am searching for would put human needs and desires back into the centre of our considerations. But it would also satisfy the requirements which the Routleys think that an environmental ethic should satisfy. It would be human centred without being "human chauvinistic". This will be accomplished if we can make a case for saying that human self-realization and happiness can only be achieved in a full sense if we acknowledge and treat natural systems and living creatures as worthy of respect for what they are — and not merely as instruments for achieving human satisfactions.

This idea is not an original one. It has been found in one form or another within the romantic tradition.[22] More recently it has been put forward in some of the writings of Herbert Marcuse.

IV

Marcuse argues for a new relationship between humanity and nature. In such a relationship "the object world would no longer be experienced in the context of aggressive acquisition, competition and defensive possession"; and, furthermore, nature would become an environment in which human beings would be "free to develop the specifically human faculties: the creative, aesthetic faculties".[23] Developing this kind of sensibility involves becoming receptive towards nature, developing the "ability to see things in their own right, to experience the joy enclosed in them, the erotic energy of nature".[24] It means conceiving of nature "as a cosmos with its own potentialities, necessities and chances".[25] Marcuse suggests that because of this capacity there is a sense in which nature can be regarded as a subject — though he does not mean by this that nature has consciousness, or a plan or purpose.

Although Marcuse talks generally of the human-nature relationship, what he says can readily be adapted by those who are concerned with defending wilderness. A wilderness is a coherent natural system with an identity of its own. It continually changes and yet it persists through change. Thus it has some of the same properties that we think of subjects as having, and as such is something that in Marcuse's view we should appreciate, respect and be receptive to. In fact, it seems more like the subject that Marcuse has in mind than nature as a whole. For a wilderness is an environment in which natural systems have been left pretty much to "form themselves in their freedom".

Marcuse does not say that we should never use nature in an instrumental way.[26] But how we use nature becomes of critical importance. In *One Dimensional Man* he calls for a new science and technology and new methods of production.[27] What he is asking for is not completely clear, but most of what he says suggests something that is not at all mysterious and unthinkable.[28]

The "new science" can be understood, in part, as the systematic knowledge of natural cycles and interrelations in ecosystems; and the "new technology" as the techniques for applying this knowledge in ways that are least disruptive to them. This is the science and and technology for people who respect and appreciate natural systems. Such people will, as much as possible, try to adjust their activities to natural cycles, using them to accomplish their purposes. People who have Marcuse's view of nature will prefer to use natural predators and complementary planting rather than monoculture and pesticides. They will prefer to develop and use the plants and animal species that are already native rather than importing seeds and stock from outside, and they will prefer to avoid certain ways of using animals and natural systems — for example, they will probably not engage in factory farming. They will have such preferences even if the more drastic kinds of interference are not harmful to human life and health.

Marcuse's view does not involve attributing to natural systems intrinsic value, as the Routleys do; for this kind of value is independent of human beings and their states of consciousness in a way that Marcuse's values are not. Marcuse thinks that we should value nature because it is good for us to do so. Nevertheless, I think that Marcuse's environmental ethic — that is, his view of the proper relationship between humans and natural systems — does all the important things that the Routleys and others want an environmental ethic to do.

It provides reasons for valuing wilderness and species and wanting to preserve them, even if human survival and health do not depend on doing so. It puts the onus of justification on those who want to destroy or alter a wilderness or on those who want to disrupt natural cycles, rather than on the preservationist, as it often is now. It gives a reason why we should sometimes sacrifice material satisfactions in order to preserve. And it accounts for why we feel that some ways of manipulating animals and natural systems are undesirable.

Further, it seems to account for responses to the Routleys' hypothetical situations. Would it be morally reprehensible for the last humans on earth to set about systematically destroying the living things of the earth? I think that someone who adopts Marcuse's view of the value of natural systems would have to say "yes, it is". For the qualities we are supposed to appreciate in an ecosystem are those qualities which make it a dynamic independent system, able to flourish and develop in its own way. If we do value these qualities then we would not care to put an end to them. In fact, if the last people have this kind of respect they will probably derive some comfort from the thought that nature will continue to blossom and flourish after they are dead. Nor will destruction of present ecosystems be all right if it is known that many years in the future the earth will again contain rich and varied natural systems. What we want to continue after we are gone are the natural environments and species that we have come to know and appreciate.

Nor is the kind of respect and appreciation for nature which Marcuse advocates something that is likely to be satisfied by the wilderness-experience machine. Nature, in his view, does not have the instrumental value of producing aesthetic pleasure or some other desirable emotion. To get the kind of fulfillment out of nature that Marcuse thinks is desirable it must be approached with the attitude that it has qualities that are worth understanding and appreciating even if some effort is required to discover them. What is supposed to be gained from that approach to nature is not a passive experience of pleasure which might be obtained from some other instrument. If to appreciate nature properly, it must be appreciated for what it is, then presumably no substitute for wilderness will be adequate.

The critical question is, how can we defend Marcuse's view of the relationship between humanity and nature? When Passmore talks about the "minor tradition", of which Marcuse is a representative, he is skeptical about this possibility. How can we insist that everyone must appreciate natural systems? "The geometrical garden has a beauty of its own . . . and so has the city of great boulevards, however unnatural the square and the straight line."[29] There are even things to appreciate about monocultured land or oil refineries. Again we come back to Passmore's original position on the psychological value of wilderness and natural systems. Some people have a taste for these things, others do not. There is no good reason why everyone should. What Passmore appears not to notice is that Marcuse's position includes a social critique and, based on this, a view about what it is to live a good life.

V

We come back again to the question: How is an ethical revolution possible? It may be true that more and more people are developing a respect and appreciation for nature. But changes in ethical consciousness, however widespread, cannot by themselves bring about a revolution in ethics. These changes can always be dismissed as the result of confusion or fashion. To show how an ethical revolution is possible we have to examine the reasons for the existence of the new consciousness.

We should begin, I suggest, by looking for developments within our cultural tradition which seem to be providing a new perception of our relation to nature, or which challenge our present ideas about the good life. We will look for the crises in social life; the criticisms that people voice about social relations and lifestyles. We cannot expect that the considerations advanced in support of a new environmental ethic will be such that any rational person will be forced to accept it. We are looking for reasons which can persuade people to see and feel differently about things. But this search is bound to take us outside the framework of ethics, as most philosophers conceive it. I want to look briefly at two considerations which lie behind the new ethical consciousness and which do seem to be capable of making a new environmental ethic possible.

The Routleys pave the way for a new environmental ethic by undermining what they call "human chauvinism" — an attitude which sanctions treating humans as a privileged species.[30] The Routleys can find no justification for this attitude and thus regard traditional human-centred ethics as depending on an unwarranted distinction between humans and nonhumans. In my view, the Routleys do not succeed in establishing that a human-centred ethic is unsupportable or undesirable. However, the term "human chauvinism" appropriately describes a still very prevalent view that humans are ontologically or metaphysically special; the belief that our destiny and what is good for us are quite different from the destiny and good of every other living thing; the assumption that we are the highlight and the culmination of life on earth. Religion has usually provided an apology for these ideas, but philosophy and even science have at various times been used to prop them up.

Those who are human chauvinists in this sense, are likely to regard things in the natural world as being important only in relation to people. They find it easy to assume that humans alone are truly valuable; and the rest of nature appears in their calculations as a hinderance or an instrument. What has made it possible to challenge this set of beliefs and assumptions are developments in science. Ways of looking at our relation to nature, which scientific theories have provided, have thus encouraged a new way of valuing.

The developments that have undermined the traditional view of our relation to nature are evolutionary theory coupled with the new science of ecology. Evolutionary theory, properly understood, does not place us at the pinnacle of the development of life on Earth. Our species is one product of evolution among many others. Like other species we will flourish only as long as our environment permits us to; and like all other species we will sooner or later become extinct, and life on Earth will continue to develop without us. Ecology complements this picture. It stresses our dependence on natural systems and our close relationships with other forms of life. We are led to see ourselves as part of an interdependent web of living things; our survival is bound up with that of the lowliest plants and animals.

The change in perception that these theories require does not necessitate a change in values. In fact, some people take it to be the message of ecology that we should be more careful in the way we use nature as an instrument. On the other hand, for many people, the change of world view and an adoption of a new way of valuing go hand in hand. As Rolston says about the influence of ecology:

For some observers at least, the sharp is/ought dichotomy is gone; the values seem to be there as soon as the facts are fully in, and both alike are properties of the system.[31]

The attempt to express the implications of an altered view of our relation to nature is common in the writings of those who have thought about the implications of ecology. The "deep" ecology movement which Arne Naess refers to is based on this new perception:

To the ecological field-worker, the equal right to live and blossom is an intuitively clear and obvious value axiom. Its restriction to humans is an anthropocentrism with detrimental effects upon the life quality of humans themselves.[32]

The language in which these environmentalists attempt to express their insight is often clumsy and misleading; the analogies they use are sometimes unfortunate; and it is usually not difficult to poke holes in their positions. But we should not lose track of the basic insight — the idea that we should see ourselves as part of a network of natural relationships and should relocate our interests and way of life accordingly.[33]

Despite the persuasive support of a scientific world view, the new ethic of the environment still seems mystical or impractical to most people who are the heirs of the Western tradition. In our world, the achievement of happiness and self-fulfillment, as we understand it, continually comes into conflict with preservation of wilderness and species. People want the roads, the electric power, the paper, etc., that seem to require the destruction of more and more wilderness. Respect and appreciation towards nature is therefore something we can only exercise in interludes between activities predicated on destruction: The logger pauses for a moment to enjoy the beauties of nature. It is not surprising that the kind of psychological value which people in our society get out of nature can be dismissed as nothing more than an agreeable feeling which some substitute can no doubt provide. Ethical revolution thus presupposes social critique: an attempt to show that present social relations, and the goals and desires that spring from them, are unsatisfactory, and that new conceptions of self-fulfillment and happiness are desirable. It is this that Marcuse, and some other radical political thinkers attempt to provide.[34]

The fulcrum of these attempts to shift the social world is a collection of views about human needs and social ideals which are, in fact, fairly common in our Western tradition. Self-fulfillment and happiness, in this view, come from being a creative, active member of your community, capable of shaping your own destiny in your social context, having a wide range of skills and knowledge which are relevant for this purpose and being able to appreciate and enjoy a wide variety of objects and activities.[35] This kind of self-fulfillment is thwarted by the domination of one group of people over another, of which there are many instances in our society; by social relations that result in work being deadening, uncreative and out of the control of those working; and by social institutions that render individuals powerless and passive. Marcuse, and others who pursue this critique of domination, find this suppression of real human needs to be part of both Western capitalism and Eastern "socialism". The satisfactions that Western society does offer its citizens — the goods and entertainment it provides to those who can afford them — do not satisfy their true needs or allow them to realize their real potential, though they do make people passively content with things as they are.[36]

I am not here going to attempt to defend these ideas about what true human needs or potentialities are. I will content myself with pointing out that there is nothing "un-Western" or even unusual about them. Nor will I attempt to further describe or defend the criticisms of Western, or Eastern, societies. What I am interested in here is how Marcuse, and others who follow the same critical path, connect their view about society with their insistence on the necessity of a new relation between humanity and nature and, in consequence, a new ethic of the environment.

Marcuse believes that domination in the social world is linked to the domination of nature in such a way that the disappearance of one form of domination requires the removal of the other. Behind this idea lurks Hegel's dialectic of the master and the slave. Marcuse's innovation is to cast nature in the role of the slave. In fact, Marcuse sees the relation of human masters to human slaves as both a special case of the mastery of nature and something which results from this mastery. Hegel's master seeks self-identity through being reflected by another, and attempts, unsatisfactorily, to achieve this by dominating another. In the same way, suggests Marcuse, the masters of nature seek to make nature reflect them — to make over nature in the human image. But this attempt to conquer and subdue nature has some undesirable consequences. First of all, because humans themselves are a part of nature, the enslavement of nature tends to lead to the enslavement of humans themselves by means of the same science and technology that is used to conquer nature. And secondly, the degradation of nature not only affects the prospects for human survival, but also diminishes human self-identity and worth.[37]

However, this connection between an exploitative approach to nature and the degradation of the human psyche is not an obvious one. Indeed most social critics, even those who share many of Marcuse's views about the true needs and potential of humans, have not seen it. Why couldn't it be possible to construct a society in which the fulfillment of true human needs was possible without making any change in our relationship with nature?

In *State and Revolution,* Lenin's social ideals do not seem at first to be appreciably different from those held by Marcuse and many anarchists. He too looks forward to a society without economic and political domination. But he assumes that this will be achieved in a highly centralized society in which sophisticated, large-scale technology can be used to do the boring, debilitating work. Administration and co-ordination of this machinery is something that can be done by anyone who has a bookkeeper's skill.[38] The work that remains will be shared by all. Otherwise people will have the leisure time to develop themselves, to participate in public affairs, discuss philosophy – to lead, in short, a life similar in many ways to that which Aristotle recommends for Athenian citizens; only now this lifestyle is predicated on the slavery of nature and mechanisms and not on the slavery of other people. Human liberation entails not a liberation of nature but a more perfect mastery over it.[39]

Marcuse's critique of domination takes up a theme which has come to the fore in the last ten years: domination, he argues, is inherent in certain kinds of technology, ways of thinking and the social relations that go with them. The point he seems to make against Lenin's vision of socialism is that the technology it requires and the organization of this technology would not really allow human liberation and personal self-realization in any true sense. It is not just that it is hard to believe that an administration of things would not become an administration of people – by the technicians and managers who understand and coordinate the technology. But that a highly centralized, automated system of production imposes its own requirements on social life. It depends, for example, on a coordinated and disciplined work force. Work for most people in such a society is unlikely to be creative; nor in practice are individuals, or even communities, likely to have much control over, or knowledge of, this technology and how it is used.

When Marcuse, in *One Dimensional Man,* insists on the need for a new technology and science, one of the points he is making is that ways of producing which have been developed in a society containing relationships of domination are not likely to be the kind appropriate for a liberated society.[40] As Marcuse sees it, the human ideals which are widespread, if not universally accepted, in our Western tradition are in conflict with our technology and social relations and therefore with some of the ethical ideas which are also part of Western tradition.

Not surprisingly, most of those who have attempted to work out the implications of this critique argue that a society in which people are able to live truly satisfying lives will have to be a decentralized society using small-scale, though perhaps sophisticated, technology; a society in which communities and individuals have a considerable amount of autonomy and in which people have decision-making power over how their material and spiritual needs are provided for, and may be engaged to a much greater extent than is possible now in producing for their own use. Not every social critic requires that we shift to rural communities and grow our own vegetables (as the Routleys recommend) but they do generally recommend some moves to do what Marx and Engels, in the *German Ideology,* said was desirable — to break down the separation between town and country.[41]

What does this have to do with preservation of wilderness and species? There are some empirical reasons for supposing that the new way of life envisioned by such social critics will also be conducive to a respect for the natural environment. If people derive their satisfaction, not from buying and consuming, but from social relations and creative activity; and if they live in such a way that it is unnecessary to use many of the goods now necessary to carry on modern life, then their impact on the environment is likely to be less. People can afford to leave wilderness alone; they are free to develop respect and appreciation for natural systems, if they feel so inclined.[42] And there are also reasons for believing that many will feel so inclined. For people who live in contact with natural processes and understand how their daily existence is dependent on them, are more likely to respect and appreciate them, to see the land and other living things as part of their community, and natural systems as "subjects". Probably it is not necessary for people to work on the land in order to have this appreciation. People who are involved in making decisions about how land should be used and how basic needs should be satisfied are going to have to learn about the ecological relationships that they depend on. I have already argued that this kind of understanding can, and often does, foster respect and appreciation of natural systems, a feeling of identification with one's environment.

It has to be admitted that none of this guarantees the preservation of wilderness and species. It may, in fact, be naively optimistic to suppose that preserving wilderness and natural systems and promoting human happiness, however conceived, can fit together so perfectly. Even under the most ideal social arrangements, conflicts between human needs and wilderness preservation are not inconceivable. Human beings are endowed by nature with certain inalienable needs — like the need for food and shelter — and if the population is great enough, these elementary needs are bound to come into conflict with the goal of preservation, however respectful people are towards nature.

The rosy vision of the good life proposed by advocates of decentralized communities may also seem too good to be true. Is this view of the good life really compatible with, or conducive to, a new environmental ethic, or the solution of environmental problems? Decentralization could instead lead to village rivalry, petty acquisitiveness, "rural idiocy" and consequent neglect of the environmental well-being. Or perhaps it would encourage the desire for large families and result in a population crisis. Perhaps some environmental and social problems can only be overcome by centralized organizations and large-scale planning. However, the purpose of this essay is not to defend any particular idea of the good life, but to try to show how social criticism underlies ethical revolution: to show how a new environmental ethic can be possible.

Notes

1. Cited in Dick Johnson, ed., *The Alps at the Crossroads* (Melbourne: Victorian National Parks Association, 1974, p. 19.
2. John Passmore, *Man's Responsibility for Nature* (London: Duckworth, 1974), chap. 4, especially pp. 109–10.
3. Passmore distinguishes between conservation – the prudent management of natural resources for the benefit of present and future generations – and preservation – retaining our environment in its original state (Passmore, *Man's Responsibility,* pp. 73 and 101).
4. Richard and Val Routley, "Human Chauvinism and Environmental Ethics", in *Environmental Philosophy,* ed. Don Mannison, Michael McRobbie and Richard Routley (Canberra: Research School of Social Sciences, Australian National University, 1980), pp. 136–37.
5. It might be objected that the wilderness-experience machine cannot be an adequate substitute for wilderness simply because what it provides is not the real thing. This does not, however, undermine the point that the Routleys are trying to make: that the mental satisfactions produced by wilderness might be produced by something else.
6. In the recent literature, John Rodman, "The Liberation of Nature", *Inquiry*

20 (1977): 83–145, argues that an environmental ethic capable of satisfying the ethical perceptions of the environmentally concerned cannot simply be an extension of traditional ethical principles and categories. Holmes Rolston III, "Is There an Ecological Ethic?", *Ethics* 85 (1974–75): 93–109, also claims that for some "the ecological perspective penetrates not only the secondary, but the primary qualities of an ethic . . . ," (i.e., it requires new first premises) (p. 98).

7. "Just as our statements about other people tend to be concealed statements about ourselves, so statements about non-human nature tend to be concealed statements about the human condition, and movements to liberate non-human nature tend also to be movements to liberate the repressed potentialities of human nature" (Rodman, "Liberation of Nature", p. 105).

8. The difficulties that stand in the way of a new environmental ethic are stressed in a number of articles in *Environmental Philosophy*, ed. Mannison et al.; see Robert Elliot, "Why Preserve Species?"; H. J. McCloskey, "Ecological Ethics and Its Justification: A Critical Appraisal"; Don Mannison, "A Critique of a Proposal for an Environmental Ethic". These difficulties are also demonstrated by those who undertake a search for a new ethic. For instance, E. P. Odum and W. T. Blackstone both hold that an ethical change is necessary and yet seem to settle for a variation of traditional ethics. See Odum, "Environmental Ethics and the Attitude Revolution" and Blackstone, "Ethics and Ecology", both in *Philosophy and Environmental Crisis*, ed. William T. Blackstone (Athens, Ga: University of Georgia Press, 1974).

9. The usual treatment of ethics also encourages the idea that ethical rules restrain us, discipline us, in an analogous way to government of the state. It does not permit us to see ethical behaviour as flowing naturally out of a particular way of life. See Rodman, "Liberation of Nature", p. 103. For a detailed criticism of "psychologised statism" see Anthony Skillen, *Ruling Institutions* (Sussex: Harvester, 1977), pp. 122–77.

10. Passmore, *Man's Responsibility*, pp. 111–17.

11. See Raymond Williams' discussions of the social ideas of some of the British romantics in *The Country and the City* (London: Chatto & Windus, 1973), Chaps. 19–21; and also in *Culture and Society 1780–1950* (London: Chatto & Windus, 1961), part I.

12. Routley and Routley, "Human Chauvinism", pp. 96–189.

13. Ibid., pp. 121–25.

14. The Routleys provide a number of variations on the "last people situation"; see Routley and Routley, "Human Chauvinism", pp. 121–23; and Richard Routley, "Is There a Need for a New, an Environmental, Ethic?", in *Proceedings of the 25th World Congress of Philosophy* (Varna, 1973): 207–8.

15. Routley and Routley, "Human Chauvinism", p. 130.

16. Ibid., pp. 124–25. Rodman, too, emphasizes that the same attitudes which condemn misuse or destruction of wilderness also condemn certain ways of manipulating animals, perhaps calling into question most forms of domestication. Rodman "Liberation of Nature", pp. 100–101.

17. Routley and Routley, "Human Chauvinism", p. 170. This value axiom might be seen as another version of Aldo Leopold's well known ethical principle: "A thing is right, when it tends to preserve the integrity, stability and beauty of the biotic community. It is wrong when it tends otherwise" (Aldo Leopold, "The Land Ethic", in *A Sand County Almanac* [New York: Oxford University Press, 1949], p. 240). However, what Leopold is actually arguing for is, I think, closer to the position advocated herein.

18. Adrian Desmond, *The Hot-Blooded Dinosaurs* (London: Blond and Briggs, 1975), p. 209.

19. Richard and Val Routley, "Social Theories, Self Management and Environmental Problems", in *Environmental Philosophy*, ed. Mannison et al., p. 326.
20. Ibid., p. 319.
21. I am not suggesting that the distinction is never useful. My point is that in a situation when values are being changed or redefined it can be confusing and artificial.
22. It is also a common theme in the recent writings of those concerned with the environment: "An enlightened anthropocentrism acknowledges that in the long run, the world's good always coincides with man's own most meaningful good. Man can manipulate nature to his best interests only if he first loves her for her own sake" (Rene Dubos, *A God Within* [London: Angus and Robertson, 1973], p. 45; cited in Rolston, "Ecological Ethic", p. 104).
23. Herbert Marcuse, "Nature and Revolution", in *Counter-Revolution and Revolt* (Boston: Beacon Press, 1972), p. 64.
24. Ibid., p. 74.
25. Ibid., p. 69.
26. Jürgen Habermas, "Technology and Science as 'Ideology' ", in his *Towards a Rational Society* (London: Heinemann, 1971), pp. 87–88, argues that Marcuse's new science and technology are ruled out because a manipulative approach to nature is a requirement of human existence. What Marcuse objects to, however, is not all of the ways in which humans use nature, but the kind of uses that can be called "manipulative" in the perjorative sense of that word.
27. "The development of the scientific concepts may be grounded in an experience of nature as a totality of life to be protected and cultivated, and technology would apply this science to the reconstruction of the environment of life" (Herbert Marcuse, *One Dimensional Man* [Boston: Beacon Press, 1964], p. 61).
28. But sometimes Marcuse identifies quantification as the source of the repressive nature of present science. Sometimes he depends on characterizations of science provided by quotes from instrumentalists, operationalists or phenomenalists. The terms "instrumentalism" and "operationalism" are then used to suggest that science is a tool for manipulating nature in a bad sort of way. These particular criticisms seem to me to rest on a misunderstanding of the nature of science.
29. Passmore, *Man's Responsibility*, p. 39.
30. Routley and Routley, "Human Chauvinism", pp. 96–115.
31. Rolston, "Ecological Ethic", p. 101.
32. Arne Naess, "The Shallow and the Deep, Long-Range Ecology Movement", *Inquiry* 16 (1973): 96.
33. Passmore dismisses such notions as mystical, the misconceived attempts to transplant the religious ideas of the East to the West (see Passmore, *Man's Responsibility*, pp. 173 ff). I have argued that they are better seen as attempts to draw out the ethical implications of our scientific knowledge. If this knowledge seems more compatible with the religion and philosophy of the East than the religion and philosophy of the West, then this is just one of those ironies of history.
34. There are a number of Marxists and anarchists whose social critiques can be seen as attempts to provide the kind of foundation for a new environmental ethic that I am advocating. They include, Murray Bookchin (Lewis Herber), *Post-Scarcity Anarchism* (Berkeley Calif.: Ramparts Press, 1971); Andre Gorz, *Ecology and Politics* (Boston: South End Press, 1980); Alan Roberts, *The Self-Managing Environment* (London: Allison & Busby, 1979).
35. For Marcuse's notion of liberation and his distinction between true needs and

false needs, see Marcuse, *One Dimensional Man*, pp. 4–8. The early Marx of *The 1844 Manuscripts* is the main inspiration for Marcuse's views about human needs.

36. This is the main theme of Marcuse, *One Dimensional Man.*
37. "Commercialised nature, polluted nature, militarized nature cut down the life environment of man, not only in an ecological but also in a very existential sense. . . . It deprives man from finding himself in nature; it also prevents him from recognizing nature as a subject in its own right – a subject with which to live in a common human universe (Marcuse, "Nature and Revolution", p. 60).
38. "We ourselves, the workers, will organize large-scale production on the basis of what capitalism has already created, relying on our own experience as workers, establishing strict, iron discipline supported by the state power of the armed workers; we will reduce the role of the state officials to that of simply carrying out our instructions as responsible, revocable, modestly paid 'foremen and bookkeepers' (of course with the aid of technicians of all sorts, types and degrees)" (V. I. Lenin, "State and Revolution", in *Selected Works,* 45 vols., [Moscow: Foreign Languages Press, 1952], vol. 2, pp. 249–50).
39. The Routleys attribute this vision of a future society to all Marxists; see Routley and Routley, "Environmental Problems", pp. 217–32. Indeed, Marx himself sometimes glorifies bourgeois mastery over nature and seems to look forward to more of the same under socialism and communism. But sometimes he also advocates breaking down the division of labour in a way that seems impossible in the kind of industrial society which now exists; see Karl Marx and Friedrick Engels, "The German Ideology: Part I in *The Marx-Engels Reader,* ed. Robert C. Tucker (New York: W. W. Norton and Ca, 1972) pp. 140–50.
40. "The point which I am trying to make is that science, by virtue of its own method and concepts, has projected and promoted a universe in which the domination of nature has remained linked to the domination of man – a link which tends to be fatal to this universe as a whole" (Marcuse, *One Dimensional Man*, p. 166).
41. Marx and Engels, "German Ideology", pp. 140–42.
42. "The Routleys develop this argument in "Environmental Problems", pp. 311–317.

**Part Two
A New Environmental Ethic?**

The Rights of the Nonhuman World
MARY ANNE WARREN

Western philosophers have typically held that human beings are the only proper objects of human moral concern. Those who speak of *duties* generally hold that we have duties only to human beings (or perhaps to God), and that our apparent duties towards animals, plants and other nonhuman entities in nature are in fact indirect duties to human beings.[1] Those who speak of moral *rights* generally ascribe such rights only to human beings.

This strictly homocentric (human-centered) view of morality is currently challenged from two seemingly disparate directions. On the one hand, environmentalists argue that because humanity is only one part of the natural world, an organic species in the total, interdependent, planetary biosystem, it is necessary for consistency to view all of the elements of that system, and not just its human elements, as worthy of moral concern in themselves, and not only because of their usefulness to us. The ecologist Aldo Leopold was one of the first and most influential exponents of the view that not only human beings, but plants, animals and natural habitats, have moral rights. We need, Leopold argued, a new ethical system that will deal with our relationships not only with other human individuals and with human society, but also with the land, and its nonhuman inhabitants. Such a "land ethic" would seek to change "the role of *Homo sapiens* from conqueror of the land community to plain member and citizen of it".[2] It would judge our interaction with the nonhuman world as "right when it tends to preserve the integrity, stability, and beauty of the biotic community", and "wrong when it tends otherwise".[3]

On the other hand, homocentric morality is attacked by the so-called animal liberationists, who have argued, at least as early as the eighteenth century (in the Western tradition), that insofar as (some) nonhuman animals are sentient beings, capable of experiencing pleasure and pain,[4] they are worthy in their own right of our moral concern.[5]

On the surface at least, the animal liberationist ethic appears to be quite different from that of ecologists such as Leopold. The land ethic is *wholistic* in its emphasis: it treats the good of the biotic *community* as the ultimate measure of the value of individual organisms or species, and of the rightness or wrongness of human actions. In contrast, the animal-liberationist ethic is largely inspired by the utilitarianism of Jeremy Bentham and John Stuart Mill.[6] The latter tradition is individualist in its moral focus, in that it treats the needs and interests of individual sentient beings as the ultimate basis for conclusions about right and wrong.

These differences in moral perspective predictably result in differences in the emphasis given to specific moral issues. Thus, environmentalists treat the protection of endangered species and habitats as matters for utmost concern, while, unlike many of the animal liberationists,[7] they generally do not object to hunting, fishing or rearing animals for food, so long as these practices do not endanger the survival of certain species or otherwise damage the natural environment. Animal liberationists, on the other hand, regard the inhumane treatment or killing of animals which are raised for meat, used in scientific experimentation and the like, as just as objectionable as the killing or mistreatment of "wild" animals.[8] They oppose such practices not only because they may sometimes lead to environmental damage, but because they cause suffering or death to sentient beings.

Contrasts such as these have led some philosophers to conclude that the theoretical foundations of the Leopoldian land ethic and those of the animal-liberationist movement are fundamentally incompatible,[9] or that there are "intractable practical differences" between them.[10] I shall argue on the contrary, that a harmonious marriage between these two approaches is possible, provided that each side is prepared to make certain compromises. In brief, the animal liberationists must recognize that although animals do have significant moral rights, these rights are not precisely the same as those of human beings; and that part of the difference is that the rights of animals may sometimes be overriden, for example, for environmental or utilitarian reasons, in situations where it would not be morally acceptable to override human rights for similar reasons. For their part, the environmentalists must recognize that while it may be acceptable, as a legal or rhetorical tactic, to speak of the rights of trees or mountains,[11] the logical foundations of such rights are quite different from those of the rights of human and other sentient beings. The issue is of enormous importance for moral philosophy, for it centres upon the theoretical basis for the the ascription of moral rights, and hence bears directly upon such disputed cases as the rights of (human) foetuses, children, the comatose, the insane, etc. Another interesting feature is the way in which utilitarians and deontologists often seem to exchange sides in the battle – the former insist upon the universal application of the principle that to cause unnecessary pain is wrong, while the latter refuse to apply that principle to other than human beings, unless there are utilitarian reasons for doing so.

In section I I will examine the primary line of argument presented by the contemporary animal-rights advocates, and suggest that their conclusions must be amended in the way mentioned above. In section II I will present two arguments for distinguishing between the rights of human beings and those of (most) nonhuman animals. In section III I will consider the animal liberationists' objection that any such distinction will endanger the rights of certain "nonparadigm" human beings, for example, infants and the mentally incapacitated. In section IV I will reply to several current objections to the attempt to found basic moral rights upon the sentience, or other psychological capacities, of the entity involved. Finally, in section V, I will examine the moral theory implicit in the land ethic, and argue that it may be formulated and put into practice in a manner which is consistent with the concerns of the animal liberationists.

I
Why (Some) Animals Have (Some) Moral Rights

Peter Singer is the best known contemporary proponent of animal liberation. Singer maintains that all sentient animals, human or otherwise, should be regarded as morally equal; that is, that their interests should be given equal consideration. He argues that sentience, the capacity to have conscious experiences such as pain or pleasure, is "the only defensible boundary of concern for the interests of others".[12] In Bentham's often-quoted words, "the question is not, Can they reason? nor, Can they talk? but Can they suffer?"[13] To suppose that the interests of animals are outside the scope of moral concern is to commit a moral fallacy analogous to sexism or racism, a fallacy which Singer calls *speciesism*. True, women and members of "minority" races are more *intelligent* than (most) animals – and almost certainly no less so than white males – but that is not the point. The point does not concern these complex capabilities at all. For, Singer says, "The claim to equality does not depend on intelligence, moral capacity, physical strength, or similar matters of fact."[14]

As a utilitarian, Singer prefers to avoid speaking of moral *rights,* at least insofar as these are construed as claims which may sometimes override purely utilitarian considerations.[15] There are, however, many other advocates of animal liberation who do maintain that animals have moral rights, rights which place limitations upon the use of utilitarian justifications for killing animals or causing them to suffer.[16] Tom Regan, for example, argues that if all or most human beings have a right to life, then so do at least some animals.[17] Regan points out that unless we hold that animals have a right to life, we may not be able to adequately support many of the conclusions that most animal liberationists think are important, for example, that it is wrong to kill animals painlessly to provide human beings with relatively trivial forms of pleasure.[18]

This disagreement between Singer and Regan demonstrates that there is no single well-defined theory of the moral status of animals which can be identified as *the* animal liberationist position. It is clear, however, that neither philosopher is committed to the claim that the moral status of animals is completely identical to that of humans. Singer points out that his basic principle of equal *consideration* does not imply identical *treatment.*[19] Regan holds only that animals have *some* of the same moral rights as do human beings, not that *all* of their rights are necessarily the same.[20]

Nevertheless, none of the animal liberationists have thus far provided a clear explanation of how and why the moral status of (most) animals differs from that of (most) human beings; and this is a point which must be clarified if their position is to be made fully persuasive. That there is such a difference seems to follow from some very strong moral intuitions which most of us share. A man who shoots squirrels for sport may or may not be acting reprehensibly; but it is difficult to believe that his actions should be placed in *exactly* the same moral category as those of a man who shoots women, or black children, for sport. So too it is doubtful that the Japanese fishermen who slaughtered dolphins because the latter were thought to be depleting the local fish populations were acting quite *as* wrongly as if they had slaughtered an equal number of their human neighbours for the same reason.

Can anything persuasive be said in support of these intuitive judgments? Or are they merely evidence of unreconstructed speciesism?

To answer these questions we must consider both certain similarities and certain differences between ourselves and other animals, and then decide which of these are relevant to the assignment of moral rights. To do this we must first ask just what it means to say than an entity possesses a certain moral right.

There are two elements of the concept of a moral right which are crucial for our present purposes. To say that an entity, X, has a moral right to Y (some activity, benefit or satisfaction) is to imply at least the following:

1. that it would be morally wrong for any moral agent to intentionally deprive X or Y without some sufficient justification;
2. that this would be wrong, at least in part, *because of the (actual or potential) harm which it would do to the interests of X.*

On this (partial) definition of a moral right, to ask whether animals have such rights is to ask whether there are some ways of treating them which are morally objectionable because of the harm done to the animals themselves, and not merely because of some *other* undesirable results, such as damaging the environment or undermining the moral character of human beings. As Regan and other animal liberationists have pointed out, the arguments for ascribing at least some moral rights to sentient nonhuman animals are very similar to the arguments for ascribing those same rights to sentient human beings.[21] If we argue that human beings have rights not to be tortured, starved or confined under inhumane conditions, it is usually by appealing to our knowledge that they will suffer in much the same ways that we would under like circumstances. A child must learn that other persons (and animals) can experience, for example, pain, fear or anger, on the one hand; pleasure or satisfaction, on the other, in order to even begin to comprehend why some ways of behaving towards them are morally preferable to others.

If these facts are morally significant in the case of human beings, it is attractive to suppose that they should have similar significance in the case of animals. Everything that we know about the behaviour, biology and neurophysiology of, for instance, nonhuman mammals, indicates that they are capable of experiencing the same basic types of physical suffering and discomfort as we are, and it is reasonable to suppose that their pleasures are equally real and approximately as various. Doubts about the sentience of other animals are no more plausible than doubts about that of other human beings. True, most animals cannot use human language to *report* that they are in pain, but the vocalizatons and "body language" through which they *express* pain, and many other psychological states, are similar enough to our own that their significance is generally clear.

But to say this is not yet to establish that animals have moral rights. We need a connecting link between the premise that certain ways of treating animals cause them to suffer, and the conclusion that such actions are *prima facie* morally wrong, that is, wrong unless proven otherwise. One way to make this connection is to hold that it is a *self-evident truth* that the unnecessary infliction of suffering upon any sentient being is wrong. Those who doubt this claim may be accused (perhaps with some justice) of lacking empathy, the ability to "feel with" other sentient beings, to comprehend the reality of their experience. It may be held that it is possible to regard the suffering of animals as morally insignificant only to the extent that one suffers from blindness to "the ontology of animal reality";[22] that is, from a failure to grasp the fact that they are centres of conscious experience, as we are.

This argument is inadequate, however, since there may be those who fully comprehend the fact that animals are sentient beings, but who still deny that their pains and pleasures have any direct moral significance. For them, a more persuasive consideration may be that our moral reasoning will gain in clarity and coherence if we recognize that the suffering of a nonhuman being is an evil of the same general sort as that of a human being. For if we do not recognize that suffering is an intrinsic evil, something which ought not to be inflicted deliberately without just cause, then we will not be able to fully understand why treating *human beings* in certain ways is immoral.

Torturing human beings, for example, is not wrong merely because it is illegal (where it is illegal), or merely because it violates some implicit agreement amongst human beings (though it may). Such legalistic or contractualistic reasons leave us in the dark as to why we *ought* to have, and enforce, laws and agreements against torture. The essential reason for regarding torture as wrong is that it *hurts,* and that people greatly prefer to avoid such pain — as do animals. I am not arguing, as does Kant, that cruelty to animals is wrong because it causes cruelty to human beings, a position which consequentalists often endorse. The point, rather, is that unless we view the deliberate infliction of needless pain as inherently wrong we will not be able to understand the moral objection to cruelty of *either* kind.

It seems we must conclude, therefore, that sentient nonhuman animals have certain basic moral rights, rights which they share with all beings that are psychologically organized around the pleasure/pain axis. Their capacity for pain gives them the right that pain not be intentionally and needlessly inflicted upon them. Their capacity for pleasure gives them the right not to be prevented from pursuing whatever pleasures and fulfillments are natural to creatures of their kind. Like human rights, the rights of animals may be overriden if there is a morally sufficient reason for doing so. What *counts* as a morally significant reason, however, may be different in the two cases.

II
Human and Animal Rights Compared

There are two dimensions in which we may find differences between the rights of human beings and those of animals. The first involves the *content* of those rights, while the second involves their strength; that is, the strength of the reasons which are required to override them.

Consider, for instance, the right to liberty. The *human* right to liberty precludes imprisonment without due process of law, even if the prison is spacious and the conditions of confinement cause no obvious physical suffering. But it is not so obviously wrong to imprison animals, especially when the area to which they are confined provides a fair approximation of the conditions of their natural habitat, and a reasonable opportunity to pursue the satisfactions natural to their kind. Such conditions, which often result in an increased lifespan, and which may exist in wildlife sanctuaries or even well-designed zoos, need not frustrate the needs or interests of animals in any significant way, and thus do not clearly violate their rights. Similarly treated human beings, on the other hand (e.g., native peoples confined to prison-like reservations), do tend to suffer from their loss of freedom. Human dignity and the fulfillment of the sorts of plans, hopes and desires which appear (thus far) to be uniquely human, require a more extensive freedom of movement than is the case with at least many nonhuman animals. Furthermore, there are aspects of human freedom, such as freedom of thought, freedom of speech and freedom of political association, which simply do not apply in the case of animals.

Thus, it seems that the human right to freedom is more extensive; that is, it precludes a wider range of specific ways of treating human beings than does the corresponding right on the part of animals. The argument cuts both ways, of course. *Some* animals, for example, great whales and migratory birds, may require at least as much physical freedom as do human beings if they are to pursue the satisfactions natural to their kind, and this fact provides a moral argument against keeping such creatures imprisoned.[23] And even chickens may suffer from the extreme and unnatural confinement to which they are subjected on modern "factory farms". Yet it seems unnecessary to claim for *most* animals a right to a freedom quite as broad as that which we claim for ourselves.

Similar points may be made with respect to the right to life. Animals, it may be argued, lack the cognitive equipment to value their lives in the way that human beings do. Ruth Cigman argues that animals have *no* right to life because death is no misfortune for them.[24] In her view, the death of an animal is not a misfortune, because animals have no desires which are *categorical*; that is which do not "merely presuppose being alive (like the desire to eat when one is hungry), but rather answer the question whether one wants to remain alive".[25] In other words, animals appear to lack the sorts of long-range hopes, plans, ambitions and the like, which give human beings such a powerful interest in continued life. Animals, it seems, take life as it comes and do not specifically desire that it go on. True, squirrels store nuts for the winter and deer run from wolves; but these may be seen as instinctive or conditioned responses to present circumstances, rather than evidence that they value life as such.

These reflections probably help to explain why the death of a sparrow seems less tragic than that of a human being. Human lives, one might say, have greater intrinsic value, because they are worth more *to their possessors*. But this does not demonstrate that no nonhuman animal has *any* right to life. Premature death may be a less *severe* misfortune for sentient nonhuman animals than for human beings, but it is a misfortune nevertheless. In the first place, it is a misfortune in that it deprives them of whatever pleasures the future might have held for them, regardless of whether or not they ever *consciously anticipated* those pleasures. The fact that they are not here afterwards, to *experience* their loss, no more shows that they have not lost anything than it does in the case of humans. In the second place, it is (possibly) a misfortune in that it frustrates whatever future-oriented desires animals *may* have, unbeknownst to us. Even now, in an age in which apes have been taught to use simplified human languages and attempts have been made to communicate with dolphins and whales, we still know very little about the operation of nonhuman minds. We know much too little to assume that nonhuman animals never consciously pursue relatively distant future goals. To the extent that they do, the question of whether such desires provide them with *reasons for living* or merely *presuppose* continued life, has no satisfactory answer, since they cannot contemplate these alternatives — or, if they can, we have no way of knowing what their conclusions are. All we know is that the more intelligent and psychologically complex an animal is, the more *likely* it is that it possesses specifically future-oriented desires, which would be frustrated even by *painless* death.

For these reasons, it is premature to conclude from the apparent intellectual inferiority of nonhuman animals that they have no right to life. A more plausible conclusion is that animals do have a right to life but that it is generally somewhat weaker than that of human beings. It is, perhaps, weak enough to enable us to justify killing animals when we have no other ways of achieving such vital goals as feeding or clothing ourselves, or obtaining knowledge which is necessary to save human lives. Weakening their right to life in this way does not render meaningless the assertion that they have such a right. For the point remains that *some* serious justification for the killing of sentient non-human animals is always necessary; they may not be killed merely to provide amusement or minor gains in convenience.

If animals' rights to liberty and life are somewhat weaker than those of human beings, may we say the same about their right to *happiness*; that is, their right not to be made to suffer needlessly or to be deprived of the pleasures natural to their kind? If so, it is not immediately clear why. There is little reason to suppose that pain or suffering are any less unpleasant for the higher animals (at least) than they are for us. Our large brains *may* cause us to experience pain more intensely than do most animals, and *probably* cause us to suffer more from the anticipation or remembrance of pain. These facts might tend to suggest that pain is, on the whole, a worse experience for us than for them. But it may also be argued that pain may be *worse* in some respects for non-human animals, who are presumably less able to distract themselves from it by thinking of something else, or to comfort themselves with the knowledge that it is temporary. Brigid Brophy points out that "pain is likely to fill the sheep's whole capacity for experience in a way it seldom does in us, whose intellect and imagination can create breaks for us in the immediacy of our sensations".[26]

The net result of such contrasting considerations is that we cannot possibly claim to know whether pain is, on the whole, worse for us than for animals, or whether their pleasures are any more or any less intense than ours. Thus, while we may justify assigning them a somewhat weaker right to life or liberty, on the grounds that they desire these goods less intensely than we do, we cannot discount their rights to freedom from needlessly inflicted pain or unnatural frustration on the same basis. There may, however, be *other* reasons for regarding all of the moral rights of animals as somewhat less stringent than the corresponding human rights.

A number of philosophers who deny that animals have moral rights point to the fact that nonhuman animals evidently lack the capacity for moral autonomy. Moral autonomy is the ability to act as a moral agent; that is, to act on the basis of an understanding of, and adherence to, moral rules or principles. H. J. McCloskey, for example, holds that "it is the capacity for moral autonomy . . . that is basic to the possibility of possessing a right".[27] McCloskey argues that it is inappropriate to ascribe moral rights to any entity which is not a moral agent, or *potentially* a moral agent, because a right is essentially an entitlement granted to a moral agent, licensing him or her to *act* in certain ways and to *demand* that other moral agents refrain from interference. For this reason, he says, "Where there is no possibility of [morally autonomous] action, potentially or actually . . . and where the being is not a member of a kind which is normally capable of [such] action, we withhold talk of rights."[28]

If moral autonomy — or being *potentially* autonomous, or a member of a kind which is *normally* capable of autonomy — is a necessary condition for having moral rights, then probably no nonhuman animal can qualify. For moral autonomy requires such probably uniquely human traits as "the capacity to be critically self-aware, manipulate concepts, use a sophisticated language, reflect, plan, deliberate, choose, and accept responsibility for acting".[29]

But why, we must ask, should the capacity for autonomy be regarded as a precondition for possessing moral rights? Autonomy is clearly crucial for the *exercise* of many human moral or legal rights, such as the right to vote or to run for public office. It is less clearly relevant, however, to the more basic human rights, such as the right to life or to freedom from unnecessary suffering. The fact that animals, like many human beings, cannot *demand* their moral rights (at least not in the words of any conventional human language) seems irrelevant. For, as Joel Feinberg points out, the interests of non-morally autonomous human beings may be defended by others, for example, in legal proceedings; and it is not clear why the interests of animals might not be represented in a similar fashion.[30]

It is implausible, therefore, to conclude that because animals lack moral autonomy they should be accorded *no moral rights whatsoever.* Nevertheless, it may be argued that the moral autonomy of (most) human beings provides a second reason, in addition to their more extensive interests and desires, for according somewhat *stronger* moral rights to human beings. The fundamental insight behind contractualist theories of morality[31] is that, for morally autonomous beings such as ourselves, there is enormous mutual advantage in the adoption of a moral system designed to protect each of us from the harms that might otherwise be visited upon us by others. Each of us ought to accept and promote such a system because, to the extent that others also accept it, we will all be safer from attack by our fellows, more likely to receive assistance when we need it, and freer to engage in individual as well as cooperative endeavours of all kinds.

Thus, it is the possibility of *reciprocity* which motivates moral agents to extend *full and equal* moral rights, in the first instance, only to other moral agents. I respect your rights to life, liberty and the pursuit of happiness in part because you are a sentient being, whose interests have intrinsic moral significance. But I respect them as *fully equal to my own* because I hope and expect that you will do the same for me. Animals, insofar as they lack the degree of rationality necessary for moral autonomy, cannot agree to respect our interests as equal in moral importance to their own, and neither do they expect or demand such respect from us. Of course, domestic animals may expect to be fed, etc. But they do not, and cannot, expect to be treated as moral equals, for they do not understand that moral concept or what it implies. Consequently, it is neither pragmatically feasible nor morally obligatory to extend to them the same *full and equal* rights which we extend to human beings.

Is this a speciesist conclusion? Defenders of a more extreme animal-rights position may point out that this argument, from the lack of moral autonomy, has exactly the same form as that which has been used for thousands of years to rationalize denying equal moral rights to women and members of "inferior" races. Aristotle, for example, argued that women and slaves are naturally subordinate beings, because they lack the capacity for moral autonomy and self-direction;[32] and contemporary versions of this argument, used to support racist or sexist conclusions, are easy to find. Are we simply repeating Aristotle's mistake, in a different context?

The reply to this objection is very simple: animals, unlike women and slaves, really *are* incapable of moral autonomy, at least to the best of our knowledge. Aristotle certainly *ought* to have known that women and slaves are capable of morally autonomous action; their capacity to use moral language alone ought to have alerted him to this likelihood. If comparable evidence exists that (some) nonhuman animals are moral agents we have not yet found it. The fact that some apes (and, possibly, some cetaceans) are capable of learning radically simplified human languages, the terms of which refer primarily to objects and events in their immediate environment, in no way demonstrates that they can understand abstract moral concepts, rules or principles, or use this understanding to regulate their own behaviour.

On the other hand, this argument implies that if we *do* discover that certain nonhuman animals are capable of moral autonomy (which is certainly not impossible), then we ought to extend full and equal moral rights to those animals. Furthermore, if we someday encounter extraterrestrial beings, or build robots, androids or supercomputers which function as self-aware moral agents, then we must extend full and equal moral rights to these as well. Being a member of the human species is not a necessary condition for the possession of full "human" rights. Whether it is nevertheless a *sufficient* condition is the question to which we now turn.

III
The Moral Rights of Nonparadigm Humans

If we are justified in ascribing somewhat different, and also somewhat stronger, moral rights to human beings than to sentient but non-morally autonomous animals, then what are we to say of the rights of human beings who happen not to be capable of moral autonomy, perhaps not even potentially? Both Singer and Regan have argued that if any of the superior intellectual capacities of normal and mature human beings are used to support a distinction between the moral status of *typical,* or paradigm, human beings, and that of animals, then consistency will require us to place certain "nonparadigm" humans, such as infants, small children and the severely retarded or incurably brain damaged, in the same inferior moral category.[33] Such a result is, of course, highly counterintuitive.

Fortunately, no such conclusion follows from the autonomy argument. There are many reasons for extending strong moral rights to nonparadigm humans; reasons which do not apply to most nonhuman animals. Infants and small children are granted strong moral rights in part because of their *potential* autonomy. But *potential* autonomy, as I have argued elsewhere,[34] is not in itself a sufficient reason for the ascription of full moral rights; if it were, then not only human foetuses (from conception onwards) but even ununited human sperm-egg pairs would have to be regarded as entities with a right to life the equivalent of our own — thus making not only abortion, but any intentional failure to procreate, the moral equivalent of murder. Those who do not find this extreme conclusion acceptable must appeal to reasons other than the *potential* moral autonomy of infants and small children to explain the strength of the latter's moral rights.

One reason for assigning strong moral rights to infants and children is that they possess not just *potential* but *partial* autonomy, and it is not clear how much of it they have at any given moment. The fact that, unlike baby chimpanzees, they are already learning the things which will enable them to *become* morally autonomous, makes it likely that their minds have more subtleties than their speech (or the lack of it) proclaims. Another reason is simply that most of us tend to place a very high value on the lives and well-being of infants. Perhaps we are to some degree "programmed" by nature to love and protect them; perhaps our reasons are somewhat egocentric; or perhaps we value them for their potential. Whatever the explanation, the fact that we do feel this way about them is in itself a valid reason for extending to them stronger moral and legal protections than we extend to nonhuman animals, even those which may have just as well or better-developed psychological capacities.[35] A third, and perhaps the most important, reason is that if we did *not* extend strong moral rights to infants, far too few of them would ever *become* responsible, morally autonomous adults; too many would be treated "like animals" (i.e., in ways that it is generally wrong to treat even animals), and would consequently become socially crippled, antisocial or just very unhappy people. If any part of our moral code is to remain intact, it seems that infants and small children *must* be protected and cared for.[36]

Analagous arguments explain why strong moral rights should also be accorded to other nonparadigm humans. The severely retarded or incurably senile, for instance, may have no potential for moral autonomy, but there are apt to be friends, relatives or other people who care what happens to them. Like children, such individuals may have more mental capacities than are readily apparent. Like children, they are more apt to achieve, or return to moral autonomy if they are valued and well cared for. Furthermore, any one of us may someday become mentally incapacitated to one degree or another, and we would all have reason to be anxious about our own futures if such incapacitation were made the basis for denying strong moral rights.[37]

There are, then, sound reasons for assigning strong moral rights even to human beings who lack the mental capacities which justify the general distinction between human and animal rights. Their rights are based not only on the value which they themselves place upon their lives and well-being, but also on the value which other human beings place upon them.

But is this a valid basis for the assignment of moral rights? Is it consistent with the definition presented earlier, according to which X may be said to have a moral right to Y only if depriving X of Y is *prima facie* wrong *because of the harm done to the interests of X*, and not merely because of any further consequences? Regan argues that we cannot justify the ascription of stronger rights to nonparadigm humans than to nonhuman animals in the way suggested, because "what underlies the ascription of rights to any given X is that X has value independently of anyone's valuing X".[38] After all, we do not speak of expensive paintings or gemstones as having rights, although many people value them and have good reasons for wanting them protected.

There is, however, a crucial difference between a rare painting and a severely retarded or senile human being; the latter not only has (or may have) value for other human beings but *also* has his or her own needs and interests. It may be this which leads us to say that such individuals have intrinsic value. The sentience of nonparadigm humans, like that of sentient nonhuman animals, gives them a place in the sphere of rights holders. So long as the moral rights of all sentient beings are given due recognition, there should be no objection to providing some of them with *additional* protections, on the basis of our interests as well as their own. Some philosophers speak of such additional protections, which are accorded to X on the basis of interests other than X's own, as *conferred* rights, in contrast to *natural* rights, which are entirely based upon the properties of X itself.[39] But such "conferred" rights are not necessarily any weaker or less binding upon moral agents than are "natural" rights. Infants, and most other nonparadigm humans have the *same* basic moral rights that the rest of us do, even though the reasons for ascribing those rights are somewhat different in the two cases.

IV
Other Objections to Animal Rights

We have already dealt with the primary objection to assigning *any* moral rights to nonhuman animals; that is, that they lack moral autonomy, and various other psychological capacities which paradigm humans possess. We have also answered the animal liberationists' primary objection to assigning somewhat *weaker,* or less-extensive rights to animals; that is, that this will force us to assign similarly inferior rights to nonparadigm humans. There are two other objections to animal rights which need to be considered. The first is that the claim that animals have a right to life, or other moral rights, has absurd consequences with respect to the natural relationships *among* animals. The second is that to accord rights to animals on the basis of their (differing degrees of) sentience will introduce intolerable difficulties and complexities into our moral reasoning.

Opponents of animal rights often accuse the animal liberationists of ignoring the realities of nature, in which many animals survive only by killing others. Callicott, for example, maintains that, whereas environmentally aware persons realize that natural predators are a vital part of the biotic community, those who believe that animals have a right to life are forced to regard all predators as "merciless, wanton, and incorrigible murderers of their fellow creatures".[40] Similarly, Ritchie asks whether, if animals have rights, we are not morally obligated to "protect the weak among them against the strong? Must we not put to death blackbirds and thrushes because they feed on worms, or (if capital punishment offends our humanitarianism) starve them slowly by permanent captivity and vegetarian diet?"[41]

Such a conclusion would of course be ridiculous, as well as wholly inconsistent with the environmental ethic. However, nothing of the sort follows from the claim that animals have moral rights. There are two independently sufficient reasons why it does not. In the first place, nonhuman predators are not moral agents, so it is absurd to think of them as wicked, or as *murdering* their prey. But this is not the most important point. Even if wolves and the like *were* moral agents, their predation would still be morally acceptable, given that they generally kill only to feed themselves, and generally do so without inflicting prolonged or unnecessary suffering. If we have the right to eat animals, in order to avoid starvation, then why shouldn't animals have the right to eat one another, for the same reason?

This conclusion is fully consistent with the lesson taught by the ecologists, that natural predation is essential to the stability of biological communities. Deer need wolves, or other predators, as much as the latter need them; without predation they become too numerous and fall victim to hunger and disease, while their overgrazing damages the entire ecosystem.[42] Too often we have learned (or failed to learn) this lesson the hard way, as when the killing of hawks and other predators produces exploding rodent populations — which must be controlled, often in ways which cause further ecological damage. The control of natural predators may *sometimes* be necessary, for example, when human pressures upon the populations of certain species become so intense that the latter cannot endure continued *natural* predation. (The controversial case of the wolves and caribou in Alaska and Canada may or may not be one of this sort.) But even in such cases it is preferable, from an environmentalist perspective, to reduce human predation enough to leave room for natural predators as well.

Another objection to assigning moral rights to sentient nonhuman animals is that it will not only complicate our own moral system, but introduce seemingly insoluble dilemmas. As Ritchie points out, "Very difficult questions of casuistry will . . . arise because of the difference in grades of sentience."[43] For instance, is it morally worse to kill and eat a dozen oysters (which are at most minimally sentient) or one (much more highly sentient) rabbit? Questions of this kind, considered in isolation from any of the practical circumstances in which they might arise, are virtually unanswerable. But this ought not to surprise us, since similarly abstract questions about the treatment of human beings are often equally unanswerable. (For instance, would it be worse to kill one child or to cause a hundred to suffer from severe malnutrition?)

The reason such questions are so difficult to answer is not just that we lack the skill and knowledge to make such precise comparisons of interpersonal or interspecies utility, but also that these questions are posed in entirely unrealistic terms. Real moral choices rarely depend entirely upon the comparison of two abstract quantities of pain or pleasure deprivation. In deciding whether to eat molluscs or mammals (or neither or both) a human society must consider *all* of the predictable consequences of each option, for example, their respective impacts on the ecology or the economy, and not merely the individual interests of the animals involved.

Of course, other things being equal, it would be morally preferable to refrain from killing *any* sentient animal. But other things are never equal. Questions about human diet involve not only the rights of individual animals, but also vital environmental and human concerns. On the one hand, as Singer points out, more people could be better fed if food suitable for human consumption were not fed to meat-producing animals.[44] On the other hand, a mass conversion of humanity to vegetarianism would represent "an increase in the efficiency of the conversion of solar energy from plant to human biomass",[45] with the likely result that the human population would continue to expand and, in the process, to cause greater environmental destruction than might occur otherwise. The issue is an enormously complex one, and cannot be solved by any simple appeal to the claim that animals have (or lack) certain moral rights.

In short, the ascription of moral rights to animals does not have the absurd or environmentally damaging consequences that some philosophers have feared. It does not require us to exterminate predatory species, or to lose ourselves in abstruse speculations about the relative degrees of sentience of different sorts of animals. It merely requires us to recognize the interests of animals as having intrinsic moral significance; as demanding some consideration, regardless of whether or not human or environmental concerns are also involved. We must now consider the question of how well the animal rights theory meshes with the environmental ethic, which treats not only animals but plants, rivers and other nonsentient elements of nature as entities which may demand moral consideration.

V
Animal Liberation and the Land Ethic

The fundamental message of Leopold's land ethic, and of the environmentalist movement in general, is that the terrestrial biosphere is an integrated whole, and that humanity is a part of that natural order, wholly dependent upon it and morally responsible for maintaining its integrity.[46] Because of the wholistic nature of biotic systems, it is impossible to determine the value of an organism simply by considering its individual moral rights: we must also consider its relationship to other parts of the system. For this reason, some philosophers have concluded that the theoretical foundations of the environmentalist and animal liberation movements are mutually contradictory.[47] Alastair Gunn states: "Environmentalism seems incompatible with the Western obsession with individualism, which leads us to resolve questions about our treatment of animals by appealing to the essentially atomistic, competitive notion of rights."[48]

As an example of the apparent clash between the land ethic and the ascription of rights to animals, Gunn points to the situation on certain islands off the coast of New Zealand, where feral goats, pigs and cats have had to be exterminated in order to protect indigenous species and habitats, which were threatened by the introduced species. "Considered purely in terms of rights," he says, "it is hard to see how this could be justified. [For,] if the goats, etc. are held to have rights, then we are violating these rights in order perhaps to save or increase a rare species."[49]

I maintain, on the contrary, that the appearance of fundamental contradiction between the land ethic and the claim that sentient non-human animals have moral rights is illusory. If we were to hold that the rights of animals are *identical to those of human beings,* then we would indeed be forced to conclude that it is wrong to eliminate harmful introduced species for the good of the indigenous ones or of the ecosystem as a whole — just as wrong as it would be to exterminate all of the human inhabitants of North America who are immigrants, however greatly this might benefit the native Americans and the natural ecology. There is no inconsistency, however, in the view that animals have a significant right to life, but one which is somewhat more easily overridden by certain kinds of utilitarian or environmental considerations than is the human right to life. On this view, it is wrong to kill animals for trivial reasons, but not wrong to do so when there is no other way of achieving a vital goal, such as the preservation of threatened species.

Another apparent point of inconsistency between the land ethic and the animal liberation movement involves the issue of whether sentience is a *necessary,* as well as *sufficient,* condition for the possession of moral rights. Animal liberationists sometimes maintain that it is, and that consequently plants, rivers, mountains and other elements of nature which are not themselves sentient (though they may *contain* sentient life forms) cannot have moral rights.[50] Environmentalists, on the other hand, sometimes argue for the ascription of moral rights to even the nonsentient elements of the biosphere.[51] Does this difference represent a genuine contradiction between the two approaches?

One argument that it does not is that the fact that a particular entity is not accorded moral rights does not imply that there are no sound reasons for protecting it from harm. Human health and survival alone requires that we place a high value on clean air, unpolluted land, water and crops, and on the maintenance of stable and diverse natural ecosystems. Furthermore, there are vital scientific, spiritual, aesthetic and recreational values associated with the conservation of the natural world, values which cannot be dismissed as luxuries which benefit only the affluent portion of humanity.[52] Once we realize how *valuable* nature is, it may seem immaterial whether or not we also wish to speak of its nonsentient elements as possessing moral *rights*.

But there is a deeper issue here than the precise definition of the term "moral rights". The issue is whether trees, rivers and the like ought to be protected *only* because of their value to us (and to other sentient animals), or whether they also have *intrinsic* value. That is, are they to be valued and protected because of what they are, or only because of what they are good for? Most environmentalists think that the natural world is intrinsically valuable, and that it is therefore wrong to wantonly destroy forests, streams, marshes and so on, even where doing so is not *obviously* inconsistent with the welfare of human beings. It is this conviction which finds expression in the claim that even nonsentient elements of nature have moral rights. Critics of the environmental movement, on the other hand, often insist that the value of the nonhuman world is purely instrumental, and that it is only sentimentalists who hold otherwise.

John Passmore, for instance, deplores "the cry ... for a new morality, a new religion, which would transform man's attitude to nature, which would lead us to believe that it is *intrinsically* wrong to destroy a species, cut down a tree, clear a wilderness.[53] Passmore refers to such a call for a nonhomocentric morality as "mystical rubbish".[54] In his view, nothing in the nonhuman world has *either* intrinsic value or moral rights. He would evidently agree with William F. Baxter, who says that "damage to penguins, or to sugar pines, or geological marvels is, without more, simply irrelevant. ... Penguins are important [only] because people enjoy seeing them walk about the rocks."[55]

This strictly instrumentalist view of the value of the nonhuman world is rejected by animal liberationists and environmentalists alike. The animal liberationists maintain that the sentience of many nonhuman animals constitutes a sufficient reason for regarding their needs and interests as worthy of our moral concern, and for assigning them certain moral rights. Sentience is, in this sense, a sufficient condition for the possession of intrinsic value. It does not follow from this that it is also a *necessary* condition for having intrinsic value. It may be a necessary condition for having individual moral *rights*; certainly it is necessary for *some* rights, such as the right not to be subjected to unnecessary pain. But there is room to argue that even though mountains and trees are not subject to pleasure or pain, and hence do not have rights of the sort we ascribe to sentient beings, nevertheless they have intrinsic value of another sort, or for another reason.

What sort of intrinsic value might they have? The environmentalists' answer is that they are valuable as organic parts of the natural whole. But this answer is incomplete, in that it does not explain why we ought to value the natural world *as a whole,* except insofar as it serves our own interests to do so. No clear and persuasive answer to this more basic question has yet been given. Perhaps, as Thomas Auxter has suggested, the answer is to be found in a teleological ethic of the same general sort of that of Plato or Aristotle, an ethic which urges us "to seek the highest good, which is generally understood as the most perfect or complete state of affairs possible".[56] This most perfect or complete state of affairs would include "a natural order which encompasses the most developed and diverse types of beings",[57] one in which "every species finds a place . . . and . . . the existence and functioning of any one species is not a threat to the existence and functioning of any other species".[58]

It is not my purpose to endorse this or any other philosophical explanation of why even the nonsentient elements of nature should be regarded as having intrinsic value. I want only to suggest that better answers to this question can and should be developed, and that there is no reason to presume that these answers will consist entirely of "mystical rubbish". Furthermore, I would suggest that the claim that mountains and forests have intrinsic value of *some* sort is intuitively much more plausible than its denial.

One way to test your own intuitions, or unformulated convictions, about this claim is to consider a hypothetical case of the following sort. Suppose that a virilent virus, developed by some unwise researcher, has escaped into the environment and will inevitably extinguish all animal life (ourselves included) within a few weeks. Suppose further that this or some other scientist has developed another virus which, if released, would destroy all plant life as well, but more slowly, such that the effects of the second virus would not be felt until after the last animal was gone. If the second virus were released *secretly,* its release would do no further damage to the well-being of any sentient creature; no one would suffer, even from the knowledge that the plant kingdom is as doomed as we are. Finally, suppose that it is known with certainty that sentient life forms would never re-evolve on the earth (this time from plants), and that no sentient aliens will ever visit the planet. The question is would it be morally preferable, in such a case, *not* to release the second virus, even secretly? If we tend to think that it would be, that it would certainly be better to allow the plants to survive us than to render the earth utterly lifeless (except perhaps for the viruses), then we do not really believe that it is only sentient — let alone only human — beings which have intrinsic value.

This being the case, it is relatively unimportant whether we say that even nonsentient natural entities may have moral *rights,* or whether we say only that, because of their intrinsic value, they ought to be protected, even at some cost to certain human interests. Nevertheless, there is an argument for preferring the latter way of speaking. It is that nonsentient entities, not being subject to pleasure or pain, and lacking any preferences with respect to what happens to them, cannot sensibly be said to have *interests.* The Gulf Stream or the south wind may have value because of their role in the natural order, but if they were to be somehow altered or destroyed, *they* would not experience suffering, or lose anything which it is in *their* interest to have. Thus, "harming" them would not be wrong *in and of itself,* but rather because of the kinds of environmental efforts which the land ethic stresses. In contrast, harm done to a sentient being has moral significance even if it has no further consequences whatsoever.

The position at which we have arrived represents a compromise between those animal liberationists who hold that only sentient beings have *either* intrinsic value or moral rights, and those environmentalists who ascribe *both* intrinsic value and moral rights to even the nonsentient elements of nature. Mountains and trees should be protected not because they have moral rights, but because they are intrinsically — as well as instrumentally — valuable.

So stated, the land ethic is fully compatible with the claim that individual sentient animals have moral rights. Indeed, the two positions are complementary; each helps to remedy some of the apparent defects of the other. The animal liberation theory, for instance, does not in itself explain why we ought to protect not only *individual* animals, but also threatened *species* of plants as well as animals. The land ethic, on the other hand, fails to explain why it is wrong to inflict needless suffering or death even upon domestic animals, which may play little or no role in the maintenance of natural ecosystems, or only a negative role. Practices such as rearing animals in conditions of severe confinement and discomfort, or subjecting them to painful experiments which have no *significant* scientific purpose, are wrong primarily because of the suffering inflicted upon individual sentient beings, and only secondarily because of any social or environmental damage they may incidentally cause.

Thus, it is clear that as we learn to extend our moral concern beyond the boundaries of our own species we shall have to take account of both the rights of individual animals *and* the value of those elements of the natural world which are not themselves sentient. Respecting the interests of creatures who, like ourselves, are subject to pleasure and pain is in no way inconsistent with valuing and protecting the richness, diversity and stability of natural ecosystems. In many cases, such as the commercial slaughter of whales, there are both environmental and humane reasons for altering current practices. In other cases, in which humane and environmental considerations appear to point in opposite directions e.g., the case of the feral goats on the New Zealand islands) these factors must be weighed against each other, much as the rights of individual human beings must often be weighed against larger social needs. In no case does a concern for the environment preclude *also* considering the rights of individual animals; it may, for instance, be possible to trap and deport the goats alive, rather than killing them.

VI
Summary and Conclusion

I have argued that the environmentalist and animal liberationist perspectives are complementary, rather than essentially competitive or mutually inconsistent approaches towards a nonhomocentric moral theory. The claim that animals have certain moral rights, by virtue of their sentience, does not negate the fact that ecosystems are complexly unified wholes, in which one element generally cannot be damaged without causing repercussions elsewhere in the system. If sentience is a necessary, as well as sufficient, condition for having moral rights, then we cannot ascribe such rights to oceans, mountains and the like; yet we have a moral obligation to protect such natural resources from excessive damage at human hands, both because of their value to us and to future generations, and because they are intrinsically valuable, as elements of the planetary biosystem. It is not necessary to choose between regarding biological communities as unified systems, analagous to organisms, and regarding them as containing many individual sentient creatures, each with its own separate needs and interests; for it is clearly both of these things at once. Only by *combining* the environmentalist and animal rights perspectives can we take account of the full range of moral considerations which ought to guide our interactions with the nonhuman world.

Notes

1. See, for instance, Immanual Kant, "Duties to Animals and Spirits", in *Lectures on Ethics,* trans. Louis Infield (New York: Harper and Row, 1964), exerpted in *Animal Rights and Human Obligations* ed. Tom Regan and Peter Singer (Englewood Cliffs, N.J.: Prentice-Hall, 1976), pp. 122–23.
2. Aldo Leopold, *A Sand County Almanac* (New York: Oxford University Press, 1949), p. 204.
3. Ibid., p. 225.
4. Here, as elsewhere in this paper, the terms "pleasure" and "pain" should not be understood in the narrow sense in which they refer only to particular sorts of *sensation,* but rather as an abbreviated way of referring to the fulfillment or frustration, respectively, of the needs, interests and desires of sentient beings.
5. See, for example, the selections by Jeremy Bentham, "A Utilitarian View"; John Stuart Mill, "A Defence of Bentham"; and Henry S. Salt, "The Humanities of Diet", "Animal Rights", and "The Logic of the Larder", in *Animal Rights,* ed. Regan and Singer.
6. Ibid.
7. See, Maureen Duffy, "Beasts for Pleasure", in *Animals, Men and Morals,* ed. Stanley and Rosalind Godlovitch (New York: Taplinger Publishing Co. 1972), pp. 111–24.
8. See, Stephen R. L. Clark, *The Moral Status of Animals* (Oxford: Clarendon Press, 1977); Tom Regan, "Animal Rights, Human Wrongs", *Environmental Ethics* 2, no. 2 (Summer 1980): 99–120; Richard Ryder, "Experiments on Animals", in *Animal Rights,* ed. Regan and Singer, pp. 33–47; and Peter Singer, *Animal Liberation: A New Ethics for Our Treatment of Animals* (New York: Avon, 1975), especially chaps. 2 and 3.
9. J. Baird Callicott, "Animal Liberation: A Triangular Affair", *Environmental Ethics* 2, no. 4 (Winter 1980): 315.
10. Ibid., p. 337.
11. See Christopher D. Stone, *Should Trees Have Standing? Toward Legal Rights for Natural Objects* (Los Altos, Calif.: William Kaufman, 1974).
12. Singer, *Animal Liberation,* p. 9.
13. Jeremy Bentham, *The Principles of Morals and Legislation* (1789), chap. 18, sec. 1; cited by Singer, *Animal Liberation,* p. 8.

14. Singer, *Animal Liberation,* p. 5.
15. Peter Singer, "The Fable of the Fox", *Ethics* 88, no. 2 (January 1978): 122.
16. See, for instance, Brigid Brophy, "In Pursuit of a Fantasy", in *Animals, Men and Morals,* pp. 125–45; Joel Feinberg, "The Rights of Animals and Unborn Generations", in *Philosophy and Environmental Crisis,* ed. William T. Blackstone (Athens, Ga. University of Georgia Press, 1974), pp. 43–68; Roslind Godlovitch, "Animals and Morals", in *Animals, Men and Morals,* pp. 156–71; Lawrence Haworth, "Rights, Wrongs and Animals", *Ethics* 88, no. 2 (January 1978): 95–105; Anthony J. Povilitis, "On Assigning Rights to Animals and Nature", *Environmental Ethics* 2 (Spring 1980): 67–71; and Tom Regan, "Do Animals Have a Right to Life?", in *Animal Rights,* ed. Regan and Singer, pp. 197–204.
17. Regan, "Right to Life?"
18. Ibid, p. 203.
19. Singer, *Animal Liberation,* p. 3.
20. Regan, "Right to Life?"; see also, idem, "An Examination and Defence of One Argument Concerning Animal Rights", *Inquiry* 22, nos. 1–2 (1979): 189–217.
21. Regan, "Right to Life?", p. 197.
22. T. L. S. Sprigge, "Metaphysics, Physicalism, and Animal Rights", *Inquiry* 22, nos. 1–2 (1979): 101.
23. See John C. Lilly, *Lilly on Dolphins* (New York: Anchor Books, Garden City, 1975), p. 210. Lilly, after years of experimenting with dolphins and attempting to communicate with them, concluded that keeping them captive was wrong because they, like us, suffer from such confinement.
24. Ruth Cigman, "Death, Misfortune, and Species Inequality", *Philosophy and Public Affairs* 10, no. 1 (Winter 1981): p. 48.
25. Ibid., pp. 57–58. The concept of a categorical desire is introduced by Bernard Williams, "The Makropoulous Case", in his *Problems of the Self* (Cambridge: Cambridge University Press), 1973.
26. Brophy, "Pursuit of Fantasy", p. 129.
27. H. J. McCloskey, "Moral Rights and Animals", *Inquiry* 22, nos. 1–2 (1979): 31.
28. Ibid., p. 29.
29. Michael Fox, "Animal Liberation: A Critique", *Ethics* 88, no. 2 (January 1978): 111.
30. Feinberg, "Rights", pp. 46–47.
31. Such as that presented by John Rawls, *A Theory of Justice* (Oxford: Oxford University Press, 1972).
32. Aristotle *Politics* 1. 1254, 1260, and 1264.
33. Singer, *Animal Liberation,* pp. 75–76; Regan, "One Argument Concerning Animal Rights".
34. Mary Anne Warren, "Do Potential People Have Moral Rights", *Canadian Journal of Philosophy* 7, no. 2 (June 1977): 275–89.
35. This argument does not, as one might suppose, justify placing restrictions upon (early) abortions which are as severe as the restrictions upon infanticide or murder, although there are certainly many people who place a high value upon the lives of foetuses. The reason it does not is that such restrictions, unlike restrictions upon infanticide (given the possibility of adoption), violate all of the most basic moral rights of women, who are not morally obligated to waive their own rights to life, liberty and happiness, in order to protect the sensibilities of human observers who are not directly affected.
36. Anthropological evidence for this claim may be found in Margaret Mead's

study of the Mundugumor, a Papuan tribe in New Guinea which placed little value on infants and abused them casually; adult Mundugumors, men and women alike, appear to be hostile, aggressive and generally amoral, to a degree barely compatible with social existence (Margaret Mead, *Sex and Temperament in Three Primitive Societies* [New York: William Morrow, 1963].).

37. One exception to the rule that mental incapacitation does not justify the denial of basic human rights is *total and permanent* incapacitation, such that there is no possibility of any future return to sentience. Once a person has entered a state of terminal coma, he or she has nothing to gain from continued biological life, and nothing to lose by dying. Where there is any doubt about the possibility of full or partial recovery, every benefit of the doubt should be given; but where there is clearly no such possibility, the best course is usually to allow death to occur naturally, provided that this is consistent with the wishes of the individual's family or friends. (To sanction *active* euthanasia, i.e., the deliberate *killing* of such terminally comatose persons might be unwise, in that it might lead all of us to fear [somewhat more] for our lives when we are forced to place them in the hands of medical personnel; but that is an issue which we need not settle here.)

38. Regan, "One Argument Concerning Animal Rights", p. 189.

39. See, for example, Edward A. Langerak, "Abortion, Potentiality, and Conferred Claims", (Paper delivered at the Eastern Division of the American Philosophical Association, December 1979).

40. Callicott, "Animal Liberation", p. 320.

41. D. G. Ritchie, "Why Animals Do Not Have Rights", in *Animal Rights*, ed. Regan and Singer, p. 183.

42. See Aldo Leopold, *Sand County Almanac*, pp. 129–33.

43. Ritchie, "Why Animals Do Not Have Rights".

44. Singer, *Animal Liberation* pp. 169–74.

45. Callicott, "Animal Liberation", p. 335.

46. For exposition of this holistic message, see, William T. Blackstone, "Ethics and Ecology", in *Philosophy and Environmental Crisis*, pp. 16–42; Thomas Auxter, "The Right Not To Be Eaten", *Inquiry* 22, nos. 1–2 (Spring 1979): 221–30; Robert Cahn, *Footprints on the Planet: The Search for an Environmental Ethic* (New York: Universe Books, 1978); Albert A. Fritsch, *Environmental Ethics* (New York: Anchor Press, Doubleday, 1980), p. 3; Alastair S. Gunn, "Why Should We Care About Rare Species", *Environmental Ethics* 2, no. 1 (Spring 1980): 17–37, Eugene P. Odum, "Environmental Ethics and the Attitude Revolution", in *Philosophy and Environmental Crisis*, pp. 10–15; and, of course, Leopold, *Sandy County Almanac*.

47. See Callicott, "Animal Liberation", p. 315.

48. Gunn, "Rare Species", p. 36.

49. Ibid., p. 37.

50. See Feinberg, "Rights", pp. 52–53.

51. See Stone, *Should Trees Have Standing?*

52. For example, Baxter maintains that "environmental amenities . . . fall in the category of a luxury good" (William F. Baxter, *People or Penguins: The Case for Optimal Pollution* [New York and London: Columbia University Press, 1974], p. 105).

53. John Passmore, *Man's Responsibility for Nature* (London: Duckworth, 1974), p. 111.

54. Ibid., p. 173.

55. Baxter, *People or Penguins*, p. 5.

56. Thomas Auxter, "The Right Not To Be Eaten", *Inquiry* 22, nos. 1–2 (1979): 222.
57. Ibid., p. 225.
58. Ibid., p. 226.

Are Values in Nature Subjective or Objective?

HOLMES ROLSTON III

I
How Should We Value Nature?

> Conceive yourself, if possible, suddenly stripped of all the emotions
> with which your world now inspires you, and try to imagine it *as
> it exists,* purely by itself, without your favourable or unfavourable,
> hopeful or apprehensive comment. It will be almost impossible for
> you to realize such a condition of negativity and deadness. No one
> portion of the universe would then have importance beyond
> another; and the whole collection of its things and series of its events
> would be without significance, character, expression, or perspective.
> Whatever of value, interest or meaning our respective worlds may
> appear embued with are thus pure gifts of the spectator's mind.[1]

William James' stark portrayal of the utterly valueless world, suddenly
transfigured as a gift of the human coming, has proved prophetic of a
dominant twentieth-century attitude. Since he wrote, we have spent
upwards of a century trying to conceive of ourselves as the sole entities
bringing value to an otherwise sterile environment. The effort has
pervaded science and technology, humanism and existentialism, ethics
and economics, metaphysics and analytic philosophy.

John Laird protested, "There is beauty . . . in sky and cloud and sea,
in lillies and in sunsets, in the glow of bracken in autumn and in the
enticing greenness of a leafy spring. Nature, indeed, is infinitely
beautiful, and she seems to wear her beauty as she wears colour or
sound. Why then should her beauty belong to us rather than to her?"[2]
But Wilhelm Windelband agreed with James: value "is never found in
the object itself as a property. It consists in a relation to an appreciating
mind, which satisfies the desires of its will or reacts in feelings of
pleasure upon the stimulation of the environment. Take away will and
feeling and there is no such thing as value."[3] R. B. Perry continued
with what became the prevailing opinion:

> The silence of the desert is without value, until some wanderer finds it lonely and terrifying; the cataract, until some human sensibility finds it sublime, or until it is harnessed to satisfy human needs. Natural substances ... are without value until a use is found for them, whereupon their value may increase to any desired degree of preciousness according to the eagerness with which they are coveted.

Any object, whatever it be, acquires value when any interest, whatever it be, is taken in it.[4]

But with the environmental turn, so surprising and pressing in the final quarter of our century, this subjectivism in values needs review. Ecology has a way of pulling into alternative focus the exchange between the organic self and the surrounding world. This can lead us to review what we have been learning in evolutionary biology and developmental biochemistry. We here argue that in the orientation of these recent sciences the subjective account of valuing becomes grossly strained. Living, as we say, "far from nature", it is remarkable to find as one of the insistent questions of our advanced civilization: how should we value nature? An ecological crisis has forced the question upon us. Environmental and evolutionary science suggest some different answers; and yet no science is quite prepared to handle the question.

Our valuational quandary is not merely a muddle into which philosophers have gotten us, although it is perhaps the last legacy of Cartesianism. Valuational incompetence is the soft underbelly of hard science. Something gone sour at the fact/value distinction is one of the roots of the ecological crisis. Values, it is typically said, form no part of nature, but only come with the human response to the world. This seems at once objective about nature and humane towards persons, but it also yields a value structure in the scientific West more anthropocentric by several orders of magnitude than were any of the value systems of the classical, Oriental, and primitive world views which have succumbed before it. But this more sophisticated view is, we think, wise in its own conceits.

The strategy in what follows is to fight a way through how we know what we know (what philosophers call epistemological issues surrounding the terms "subjective" and "objective") in order to reach the state of affairs in the real world and to be able to defend the existence of value there (what philosophers call ontological issues surrounding subjectivity and objectivity). This I do keeping the whole discussion as close to science as I can, while demanding a full-blooded, no-nonsense account of the phenomenon of value in, and valuing of, the natural world. Earlier on, I will be admitting to some inescapable blending of the subjective and objective, but later on, after this admission, I will defend all the objectivity I can for natural value.

II
Primary, Secondary, and Tertiary Qualities

Galileo's astronomy forced us to convert from a literal to a perspectival understanding of the claim that the sun is setting. His physics gave us the distinction, elaborated by Locke, between primary and secondary qualities. A secondary quality is observer dependent, manufactured out of the primary motions of matter. Colour is an experiential conversion of photon radiation; taste and smell are molecular operations. This account was problematic philosophically (as Berkeley quickly saw), but it nevertheless became entrenched. The colours and sounds which Laird found nature to wear seemed rather to go with the beholding of it, reducible in the world stripped of a perceiver to matter in motion. Coached by these theories, what was then to be said of value? If the sunset is not literally a setting sun, not even red, then surely it is not literally beautiful. Samuel Alexander proposed that values were tertiary qualities.[5] Humans agree about redness, owing to their having the same organs (apart from colour blindness, etc.), but value appraisals require an interpretive judgment twice removed from the qualities actually there.

By this account, we have no organs to taste, touch, see or smell value. So it must originate at a deeper mental level. We have no options in judging length or redness (although there are aberrations). Such experiences happen to us without any liberty to refuse them. The primary and secondary qualities are always there in the scope of consciousness. They perhaps fall into the background, but they never turn off during perception. Value judgments, by contrast, have to be decided. Beauty or utility are things we must attend to. When our minds turn aside to other thoughts, though still perceiving the object, such values entirely disappear from consciousness. We can use instruments — metre sticks, spectroscopes, thermometers, mass spectrometers — on primary and secondary qualities. Something there leaves records during photography, electrophoresis or chromatography, puzzle though we may about the experiential translation of 6800 Angstrom units into redness, or how the shape of the fructose molecule, interlocked with receptors on the tongue, is experienced as sweetness. Both primary and secondary qualities are in this sense empirical, or natural. But valuational qualities do not show up on any instruments or organs devised or conceivable. This leads some, who still look for properties in the object, to think of value as an objective but "non-natural", that is, "nonempirical" quality. But, finding nothing that produces consensus or proves researchable, most judges become convinced that these tertiary qualities are overlays, not really there in the natural world. Rainbowlike, only more so, they are gifts of the spectator's mind.

But now the puzzle deepens. Just as philosophers were reaching this consensus, a revolution in physics threw overboard the primary, supposedly objective, qualities as well. Einstein showed length, mass, time and motion to be observer dependent. They too are matters of perspective, although not so much of decision as of bodily relations. At the microlevel, Heisenberg's uncertainty principle forbade any precise hold on momentum and location. Quantum mechanics left a nonsubstantial nonpicture of nature as a gauzy haze of interpenetrating wave fields, where none of the commonsense qualities made much sense, and where even space and time grew vague. It was alarming to learn how much our mental constructions enter into the descriptions of physical science, how much the observer influences the natural phenomenon by instrumentation or sense modality. As we go down smaller, objectivity decreases more and more.

Summing this up, Einstein remarked that he had taken "the last remainder of physical objectivity" from the concepts of space and time.[6] John Wheeler wrote, "A much more drastic conclusion emerges: ... *there is no such thing as spacetime in the real world of quantum physics. . . .* It is an approximation idea, an extremely good approximation under most circumstances, but always only an approximation."[7] Werner Heisenberg wrote, "When we speak of a picture of nature provided us by contemporary exact science, we do not actually mean any longer a picture of nature, but rather a picture of our relation to nature. . . . Science no longer is in the position of observer of nature, but rather recognizes itself as part of the interplay between man and nature."[8] Before this heady sort of runaway relativism, lost in a great and unspeakable plasma, the question of values being objectively there hardly seems discussable. The subjectivists have won all the chips. Even physics, that bedrock science which gave great promise of "telling it like it is", has withdrawn entirely from that kind of claim. What hope is there for value theory to do anything more than to record what appears and seems? Any bolder claim is primitive naivete.

But when we regain our wits, such relativism can be kept under more logical control. Contary to first appearances, it can even support certain objective aspects in value judgments. For all that we have yet said, there is just as little or as much reason to think that physics is objective, as that value theory is. Judgments about what *is* (mass, space, colour) have proved observer dependent and indistinguishable from judgments about what *is good* (pleasure, beauty, grandeur). Subjectivity has eaten up everything, even the fact/value distinction. But as a matter of fact, unless we are insane, we all believe that we know some nonsubjective things about the physical world. Our judgments are not free of perceptual modification of the incoming signals, but neither, *pace* Einstein, do they lack a very large "remainder of physical objectivity". At this everyday level we do stand in some picturing relation to nature. The key is provided in Wheeler's qualification of "no such thing" with "an extremely good approximation".

"There is a hawk in the spruce beside that granite boulder." This judgment fades out on subatomic scales, diffusing away if we migrate far enough from our native range. Through optical microscopy we take photographs in colour of garnet crystals from the granite. But on the shrinking scale of electron microscopy we begin to remind ourselves that the colour (despite the black and white micrograph!) is no longer relevant, while the length and shape in crystal lattices is still pictorial. Smaller still, we become aware that we have only models of the electrons and protons which compose the granite. Shape and location dissolve into cloudy wave fields. We allow too that the weight and shape of the boulder would appear differently to an observer passing at nearly the speed of light. Even the data from the other physical levels shows up objectively, though. The spacetime dilations affect clocks, cameras, meters, one's body, and everything that ages. The nonspecifiable fuzziness of electron momentum and location registers in the bands on paper recording charts. These things are not entirely inventions of the mind, although they reveal our perspectival and theoretical reach to the microscopic and astronomical levels.

However, if I restrict the scope of my claims, none of this affects the fact that we know something objectively, factually, about hawks, spruce trees and boulders. We do not know entirely all that is there at every level, nor with an objectivity that is free from subjective contribution. But agnosticism and relativism about the ultimate structure of matter does not prevent objective knowing in a middle-level sense. The breakdown of a concept, claim or function when extrapolated does not prevent its being quite true in the restricted range that it well serves. Our partial knowing need not be illusory or false, although it is approximate and perspectival. Here value judgments too can be short-scope claims about what is the case in the mundane world. The clue provided by Alexander's word "tertiary" is not something about twice-compounded observer dependence. It is about participation at the middle-structural levels where we live. The ownership feature in value judgments is important, but we need to think of value judgments as genuine, involved (if limited) claims about the world. Afterwards, we can inquire how far they can be pressed away from our native range. They do not attach to bare primary or secondary levels, but to high-level constructions of matter with which we are in exchange — initially in common experience and afterwards in the sciences of natural history. Just as we are getting incoming commands from "out there" about length, colour, hawks and trees, so too we are getting some commands about value. We start with these as native-range judgments, not as absolute ones. They are phenomenal claims, not noumenal ones. This much makes them locally objective, although it leaves unresolved how deep they run.

III
Judgments about Types, Functions and Values

A dose of candour from the biological sciences can help cure us of the effects of the dizzying revelations of physics. Notice, for instance, that before the panorama of an ecosystem, the primary, secondary, tertiary distinction cannot do much explanatory work. If someone asks whether a thing is alive, whether it is a seed, a moss, or a microbe, whether it is edible, and tries to answer with the vocabulary of primary, secondary and tertiary qualities, however, much compounded, he can only stammer. In order to get at the richness of the natural world, we need to make many judgments for which we have directly no organs and can make no instruments, judgments to which we must attend by decision and interpretation. Here, most of our scientific judgments are third and higher order, but we nevertheless believe that through them we are accurately corresponding with the natural world. When we pass to judgments of value, we do not need to consider them radically different in kind. This erodes the dogma that factual judgments are objective while value judgments are subjective.

The *Picea* (spruce) and *Buteo* (hawk) genetic sets, for instance, are full of information. The information is long lived, reproducing itself by means of amino-acid replacement across millions of years, a kind of fire which outlasts the sticks that feed it. But is this a compounded primary, secondary, or tertiary quality? The self-maintaining know-how is there independently of our observation, unmodified by our sense perception, primary in Locke's sense. It is nonsubjective and non-secondary. It is quite as real as atoms, if also a bit nonsubstantial and fluid. To deny reality to this information on the basis of anything learned in physics is like denying that a newspaper picture contains information, because, under the lens, it turns out to be nothing but black dots. Objectively encapsulated in informational molecules, the spruce and hawk have a technique for making a way through the terrain they inhabit, pragmatic facts for their life projects. Yet these DNA based facts are not aggregated primary, secondary or even tertiary qualities, but involve advanced, emergent compositional levels.

Meanwhile, we humans who make these judgments begin with at-hand, uncontested experience and, via science, move from our native experiential level to elaborate, often unsettled theories about structural levels and their histories. We say that the genetic information has accumulated in stages. Some of the earliest information was the code for glycolysis, which evolved three and a half billion years ago, before there was atmospheric oxygen. The citric-acid cycle came later, cashing in on eighteen times as much energy as did glycolysis. Somewhere, photosynthesis evolved so that *Picea* can capture directly the energy of sunlight, with oxidative phosphorylation subsequently arising to use the atmosphere as an electron sink, improving the efficiency of the citric-acid cycle. The spruce and the hawk evolved under the prolife pressures of a selective system operating over genetic mutations, fitting them into an ecological community. In all this, we are making highly educated guesses describing the objective facts, estimates which will be partly revised, partly conserved as science advances. But most of us do not believe that we inevitably become less objective and more subjective, less primary and more secondary or tertiary as we do this. World building does go on in the mind of the beholder, as we shape up theories over experience. But world building also takes place out there. We find the information or energy flow only by attending with deliberate focus of mind. But the mind does not contribute these features because it must model them by careful attention and decision. To the contrary, we discover richer qualities in nature.

What happens now if we introduce some value judgments? We might speak of the value of nutrients, of food pyramids, of the information keying glycolysis and photosynthesis, of the exploratory value of mutations with the "good" ones conserved because they have survival value. We might speak objectively of the value of the hawk's protective colouration (even admitting the secondary nature of colour). The word *value* easily attaches to life functions as these are known at and theorized for the middle ranges of experience. We need not yet speak of *human* values, not even of experienced values, but some notion of pre-subjective value seems to belong to these "going concerns" called living organisms as they move through the environment. Value here attaches to a whole form of life and is not just resident in the detached parts as elementary units. It overleaps although it is instantiated in the individual. It appears in a holistic crossplay where neutral, lesser-valued and even disvalued parts may assume transformed value in a larger matrix. Value emerges in pronounced forms at advanced structural levels and may not be visible as a Lockean primary or secondary quality.

But, as a "tertiary" quality, value can be embedded with the facts, quite as real as the information organisms contain, sometimes just the same thing differently described. Some will object that biological "value" ought to be kept in scare quotes, since this is not what we mean by value as a quality in experienced life. One then has to trace descriptively each of the natural selections culminating in the central nervous system. We can rejoice that value emerges epiphenomenally at the very last in consciousness, but we must judge value to be absent from all the incubating steps. In all the precedents we should speak more carefully, using the term "biofunction" instead.

The haemoglobin molecule is structurally evolved from and much advanced over the myoglobin molecule. It is very much "better" (= more functional) at oxygen transport, having allosteric properties which make it a sort of microcomputer in its capacities to respond to the oxygen-exchange needs of the blooded organism, as with the hawk in flight. Lubert Stryer, a biochemist, says of it emphatically, *"In the step from myoglobin to hemoglobin, we see the emergence of a macromolecule capable of perceiving information from its environment."*[9] But a cautious value theorist will warn this chemist not to attach any "importance" to this, not to say that this step is of any value. To be really hardnosed here, this "information perceiving" is subjective poetry, only a read-back from our own experience. One might allow that it was "interesting", but not that any "interests" of the life forms in which haemoglobin evolved were at stake.

But all this careful reservation of value as a gift of the spectator's mind now seems arbitrary and narrow. As soon as we have described haemoglobin evolution, we are ready to judge it a vital and valuable upstep in the advance of life. The phenomenon of things "being important" does not arise with our awareness; it is steadily there in quantized discoveries all along the way. Galileo and Locke first subverted value theory with their mechanistic reduction of secondary to primary qualities, leaving us only an objective matter in motion. It was then compoundly subverted by Einstein's relativity, by quantum mechanics, indeterminancy and nonpicturability. These sciences probe towards ultimacy with genius and sceptical rigour. They work in the substrata with simplicity, and so leave out all but the thousandth part of an historical eventfulness that we daily experience and that other sciences do teach us to appreciate. Only by the sort of gestalt switch which can be provided by sciences at the other end of the spectrum, such as evolutionary biochemistry or ecology, dealing with the richness of natural history, can we begin to get value theory recovered from its failure of nerve.

All judgments mix theory with fact. Even the simple cases close at hand involve elements of linguistic and conceptual decision about what to call what, and where to draw the lines. An Iroquois Indian might view the hawk as his totem, or the tree and boulder as the haunt of a spirit. Certainly the scientific judgments about natural kinds (granite, *Picea, Buteo*) are theory laden. It is admittedly difficult, as philosophers of science know, to say why we prefer science to superstition, but it has to do, at least in part, with our persuasion that the one is a better window into the way things are. The interpreter imports something of himself into the interpreted. But the fact that we use theory laden decisions about natural operations does not stand in the way of description; it rather makes it possible. To know things as they objectively are, without observer bias, is a celebrated but elusive goal of natural science, a goal impossible to fully attain, but towards which we make progress. A physicist estimates the mass of a boulder, a mineralogist knows its composition; a biologist distinguishes spruce from fir; an ecologist describes an ecosystem. All are aided, not confused by their theories. "Mass", "granite", *"Picea"* and "homeostasis" are technical terms which serve as descriptive forms. The mind answers to its object of study; with progressive reformation we more approximately understand what is there.

In this context, judgments of natural value hardly differ from judgments of natural fact. In one sense, the subjectivists are in full command again and can insist that none of our seemingly objective seeing is done without wearing cultural eyeglasses. But, as before, the objectivist can reassert the common world of experience and the impressive observational force of science. We all do believe that in our native ranges humans know something of the structural levels of nature. We believe that scientific progress gives further, if approximate, access into what these natural types and processes are like. We judge between science and folklore, between good and bad science. When we then pass to judge whether this natural kind is good, or that life process has value, we are merely continuing the effort to map reality. One has to decide whether this is a *Picea*, as one has to decide whether this is a *lovely Picea*. On occasion, the judgment about value may be easier than is the judgment about fact. One can need more theory to "see" the information and energy flow, or the phylogenetic relationships, than one does to "see" the utility or beauty. That such interpretive judgments are subject to revision does not mean that value, in distinction from other natural properties, lies only in the mental state and is not an event in the spacetime track. The constructions we see always depend upon the instructions with which we look, yet the evolving mind is also controlled by the matter it seeks to investigate. This is true alike in science and in valuation.

We can be thrilled by a hawk in the windswept sky, by the rings of Saturn, the falls of Yosemite. We can admire the internal symmetry of a garnet crystal or appreciate the complexity of the forest humus. All these experiences come mediated by our cultural education; some are made possible by science. An Iroquois Indian would have variant experiences, or none at all. But these experiences have high elements of giveness, of finding something thrown at us, of successful observation. The "work" of observation is in order to better understand. In value theory too we have as much reason to think that our appreciative apparatus is sometimes facilitating, not preventing, getting to know what is really there.

Some natural values are of the commonsense kind and nearly universal to cultures, as with the taste of an apple, the pleasant warmth of the spring sun, the striking colours of the autumn. Even though these experiences come culturally bound, some natural impact here is shared by Iroquois and Nobel prizewinner. Experience is required, but something is there which one is fitted for and fitting into; some good is transmitted and is productive of the experience. The native enjoying, just because it is relational to nature but so universal among humans, faithfully attests what is there. It is precisely our experiential position as humans-in-nature which gives us factual access to events. Other natural values are opened up to us by scientific culture, by lenses and experiments. It is precisely our advanced knowledge, setting us apart from nature in theoretical abstraction over it, which takes us deeper into nature. Sometimes the purest revelations of science put us in a better position to evaluate these things as they objectively are.

IV
Natural Valuing as Ecological-Relational

We next present an explanation sketch of valuing consistent with natural history. Our inquiry is about the kind of natural value met with in unlaboured contexts, as in pure rather than applied science, in contemplative outdoor recreation rather than in industry, in ecology rather than in economics. We are not considering, for instance, how molybdenum has value as an alloy of steel, a use which it does not have in spontaneous nature. Further, we should be cautioned against thinking that nature has some few kinds of value, or no disvalue. Nature is a plural system with values unevenly distributed and counterthrusting. Like the meanings in life, values too may come piecemeal and occasionally. Still, they come regularly enough for us to wonder whether we are coping with some value tending in the system.

Consider a causal sequence (A, B, C, D) leading to the production of an event associated with natural value (E_{nv}) which produces an event of experienced value (E_{xv}), perhaps of the beauty in a waterfall or the wealth of life in a tidal zone (see figure 1). Our consciousness also

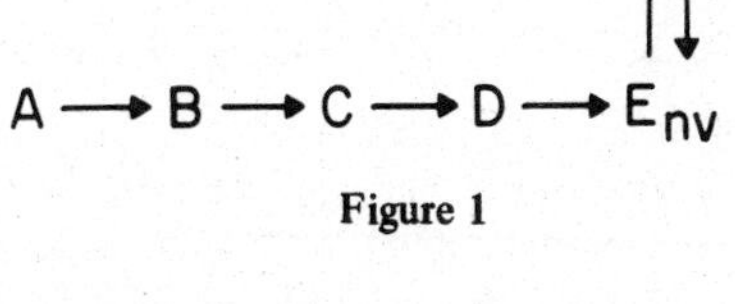

$$E_{xv}$$
$$\uparrow\downarrow$$
$$A \longrightarrow B \longrightarrow C \longrightarrow D \longrightarrow E_{nv}$$

Figure 1

responds to the waterfall or estuary, so that we need a reverse arrow (↓) making the affair relational. One is first tempted to say that value does not lie in either polar part, but is generated in their relations. Like science or recreation, valuing before nature is an interactive affair.

However, with this, more has been reallocated to the natural world than may at first be recognized. The act of responding has been ecologically grounded. We pass from abstract, reductionist, analytic knowledge to a participant, holistic, synthetic account of humans-in-nature. The subjective self is not a polar opposite to objective nature, not in the dyadic relation suggested by the paired arrows. It is rather enclosed by its environment, so that the self does value in environmental exchange in the manner indicated in figure 1, and on countless occasions. This is more-accurately represented (by E'_{nv}, E''_{nv}), in figure 2. The self has a semipermeable membrane.

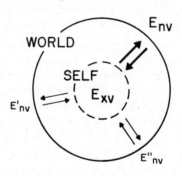

Figure 2

The setting is a given fact, a datum of nature, even though the subject must respond in imaginatively resourceful ways. I see things out there in the "field" which I choose to value or disvalue. But on deeper examination I find myself, a valuing agent, located within that circumscribing field. I do not have the valued object in "my field", but find myself emplaced in a concentric field for valuing. The whole possibility is among natural events, including the openness in my appraising. John Dewey remarked that "experience is *of* as well as *in* nature".[10] We say that valuing is *in* as well as *of* nature. What seems a dialectical relationship is an ecological one. We must add notice of how the whole happening, subject and its valued object, occurs in a natural ambiance (see figure 3).

When an ecologist remarks, "There goes a badger", he thinks not

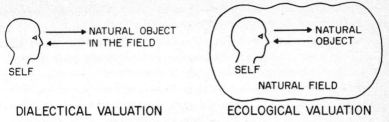

DIALECTICAL VALUATION   ECOLOGICAL VALUATION

Figure 3

merely of morphology, as might a skin-in taxonomist. He has in mind a whole mosaic of functions, interconnections, food chains; a way of being embedded in a niche where the badger is what it is environmentally. When a sociologist remarks, "There goes a vicar", he is not so much identifying a human as seeing a role in the community. The being of a vicar, like that of a badger, is a contextual affair. When a philosopher says, "There goes a valuer", he should not think of a happening inside the human in such a way as to forget that this is also an ecological event. The responsibility here is a response in our natural setting.

Add the fact that the valuing subject has itself evolved out of these surroundings. All the organs and feelings mediating value – body, senses, hands, brain, will, emotion – are natural products. Nature has thrown forward the subjective experiencer quite as much as that world which is objectively experienced. On the route behind us, at least, nature has been a personifying system. We are where this track has been heading, we are perhaps its head, but we are in some sense its tail. We next sketch a further productive sequence which generates the self (S) out of ancestral precedents (O, P, Q, R), natural events in causal sequence, and here also place reverse, valuational arrows (←) indicating reactive elements which cultural and personal responses superadd to the natural basis of personality. We add an evolutionary time line to the holistic, ecological sketch (see figure 4).

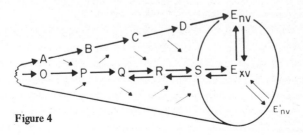

Figure 4

Seen in broad historical scale, these lines go back to common beginnings, from which they become richer, eventually to reach the experiencing self embraced by its environment. Diverse, simple and complex forms are all maintained in and by the ecosystemic pyramid, and there are many coordinating connections which we only suggest ($\searrow$). In such a picture, even though keeping the phenomenon of human valuing central, it is increasingly difficult to see valuing as isolated or even as dialectic. Values do not exist in a natural void, but rather in a natural womb.

The sudden switch in figure 1 from horizontal, merely causal arrows ($\rightarrow$), to a vertical, valuational arrow ($\uparrow$) now seems too angular a contrast. How far experienced value is a novel emergent we need yet to inquire, but there has been the historical build up towards value, and there is presently surrounding us the invitation to value. Figure 5 gives a better representation of the first series.

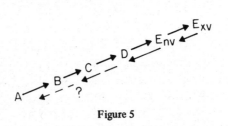

Figure 5

The reason for the new sketch is that it is difficult to say why the arrows of valuational response should value only the immediately productive natural event and not include at least some of the precedents, with unshown coordinates as well. The last event is presently at hand, and we may have had no consciousness of value during former events. But in an evolutionary ecosystem nothing happens once and *per se*; everything is embedded in a developing process.

A critic will complain, and perhaps fiercely, that we have diagramatically sketched out single sweep lines while the real world is a much more tumultous affair, where the valuational and constructive lines are not vectors but a near chaos of causes and happenstance, luck and struggle, serendipity and emergence, with much waste and little worth. The diagram screens off the heterogenous and alien character of the ingredients of value. We have straightened out strands which do not lie straight in the actual world, as though we had never heard of Darwin and his junglelike world.

There is truth in the complaint. We may wish conservatively to keep our judgments as short-scope claims. Values immediately experienced might run back to some nonvaluable base out of which they have emerged. Analogously, living organisms once emerged out of lifeless nature. A present good might have come out of historically mixed values and disvalues, as when a little good comes from much evil. Natural values might be oddly occasional, though the causal sequence is continuous. Nature is not homogenized but unevenly located, and so too with its values.

But, meanwhile, value is sometimes there before us, strikingly so, and we will sometimes be valuing contributors towards value, past or present, seen at whatever level. If ever we do extrapolate to try a systemic overview of what is going on, the likeliest account will find some programmatic evolution towards value, and this, not because it ignores Darwin, but because it heeds his principle of natural selection, and deploys this into a selection upslope towards higher values, at least along some trends within natural operations. How do we humans come to be charged up with values, if there was and is nothing in nature charging us up so? We prefer not to believe in the special creation of values, nor in their dumbfounding epigenesis — we let them evolve. Nor is our account merely a selection from the chaotic data of nature; rather, our interpretation notices how there is a world selection of events over evolutionary time (without denying other neutral or disvalued events) which builds towards the ecological valuing in which we now participate.

We can now view primary and secondary qualities holistically from above, drawn into an ecosystem at much higher structural levels, rather than viewing the ecosystem reductionistically from below, as being merely aggregated lesser qualities. We have, so to speak, an ecology of atoms and molecules. These are not described as microparticles *per se* but as events in their neighbourhoods, valued in macroscopic waterfalls and tidal basins. Genetics and biochemistry are drawn into the drama of natural history.

Many evolutionary and ecological connections are shared between ourselves as experiencers and the natural events we appraise. These bring a new orientation towards the presence of photosynthesis, the appearance of haemoglobin, or the genetic keying of information. We discover that decomposers and predators have value objectively in the ecosystem, and then realize that our own standing as subjective valuers atop the biotic pyramid is impossible, except in consequence of decomposition and predation. An interlocking kinship suggests that values are not merely in the mind but at hand in the world. We start out valuing nature like land appraisers figuring out what it is worth to us, only to discover that we are part and parcel of this nature we appraise. The earthen landscape has upraised this landscape appraiser. We do not simply bestow value on nature; nature also conveys value to us.

V
Natural Value and Consciousness

If the experience of valuing is relational, what do we say of the product, value? We must clarify the connection between experience and its objective base, since, under prevailing theories, it is widely held that the phrase "unexperienced value" is a contradiction in terms, with "experienced value" a tautology. This assumption fits the existential notice that value is not received as the conclusion of an argument, or by the indifferent observation of a causal series. A value or disvalue is whatever has got some bite to it. In the case of bare knowing, the knower has an internal *representation* of what is there, perhaps calmly so. Valuing requires more an internal *excitation.* That brings emoting, and perhaps this marriage of a subject to its object gives birth to value. It enters and exits with awareness.

Of course, if natural things have values, we cannot conceivably learn this without experiences by which we are let in on them. With every such sharing there comes a caring, and this may seem to proscribe objective neutrality. But it only prescribes circumspect inquiry. All natural science is built on the experience of nature, but this does not entail that its descriptions, its "facts", only consist in those experiences. All valuing of nature is built on experience, too, but that does not entail that its descriptions, its "values", only consist in those experiences. Valuing could be a further, non-neutral way of knowing about the world. We might suppose that value is not empirical, since we have no organs and can make no instruments for it. But it could just as well be an advanced kind of experience where a more sophisticated, living instrument is required to register natural properties. Value must be lived through, *experienced,* but so as to discern the character of the surroundings one is living through.

We next work towards this conclusion by finding inadequate some lesser accounts:

1. *Natural value as an epiphenomenon.* Pollen is not an allergen "by nature", for nasal irritation is no part of its reproductive role. But certain pollens "by accident" evoke mistakes in susceptible immunological systems. The allergy reaction is thus a disvalue which bears no meaningful connection with the natural operation. Analogously, some natural events can be (to coin a term) "valugens", evoking positive responses without meaningful basis in spontaneous nature. We react with a sense of beauty before the swirled flow in a pegmatite exposed in a rock cut. Again, we are enchanted by the mist sweeping in over the dell. But this is a kind of mis-taking of what is essentially there. Value is adventitious to nature; more fiction than fact; more dream than description; poetry, not prose; real in consciousness, unreal in the world.

But while partially useful, this account, if taken for the whole, leaves the human-valuing subject eccentric to his world. Causal connections obtain and the relational context is required, but value is a fluke without intelligible support in its stimulus.[11] In causal cases, one may be content with any kind of *how* explanation which hooks up antecedent and subsequent events. But in the case of value one would hope for an explanation more or less logically adequate to the effect. Yet so far from enlightening us about *why* value appears, this is in fact a nonexplanation. Value is an epigenetic anomaly.

2. *Natural value as an echo.* Strolling on the beach, I examine dozens of pieces of driftwood, discarding all but one. This I varnish and frame for its pleasing curvatures. I value this piece because it happened to mirror the sweep and line of my subjective preference; the rest did not. Nature once in a while chances to echo my tastes. We still have an element of accident, but we can make more sense of origins. Value does not come in pleasantly allergic reaction, but rather as a reflection of my own composition. This led Samuel Alexander to claim here that we, not nature, are the artists.

> The nature we find beautiful is not bare nature as she exists apart from us but nature as seen by the artistic eye. . . . We find nature beautiful not because she is beautiful herself but because we select from nature and combine, as the artist does more plainly when he works with pigments. . . . Nature does live for herself without us to share her life. But she is not beautiful without us to unpiece her and repiece. . . . Small wonder that we do not know that we are artists unawares. For the appreciation of nature's beauty is unreflective; and even when we reflect, it is not so easy to recognize that the

beauty of a sunset or a pure color is a construction on our part and an interpretation.[12]

But the more we reflect the less easy it becomes to see value as nothing but a reflection. Perhaps this is sometimes so but, as a general theoretical account, we have to reckon with the felicitous echoing capacity of nature, with its stimulus and surprise. Both the epiphenomenon and echo models are unecological, not sufficiently interactive and functional. To say that when enjoying blackberries or the spring sun we are participating in anomalous value seems biologically odd. The cardinal on the wing and the *Trillium* in bloom have grace, colouration and symmetry, which are structurally related to flight, flowering, and life cycles. Is the beauty here only by our selecting, not in our sensing valuative overtones that go with biological function? If I am choosing shells rather than driftwood, the colour, sweep and vault is better realized in one than in another, but seems a nisus in them all. Each attempt is an architecture under genetic control.

We are endowed with naturally selected capacities to value such things. Is the whole evolution of valuing an irrational, serendipitous afterglow? Perhaps the immunlogical system makes mistakes, but its development is incredible, except as a protection against hurts in the world. The valuational system may have fortuitous benefits, but its presence can best be accounted for in terms of an inclusive fitness to help in the world. It would be an odd benefit indeed if it did not really better fit us to our home niche. The echoing is most often working the other way around, the human valuer is reflecting what is actually there.

3. *Natural value as an emergent.* Emergent phenomena occur strikingly in nature, as when first life and afterwards learning appeared where none before existed. Perhaps the valuing capacity emerges to create value out of mere potential? Value is a kind of fiery excitement, and no natural scene, however complex and splendid, can have value until the precursors of value are supplemented and thickened by the arrival of human interest. Value cannot be said to have happened until present as an event in consciousness. There must be the delivery of some kind of "charge" into the valued experience. We may have little sense of manufacture or decision, but still we furnish the required awareness. Like knowing, the process of valuing goes on in the conscious mind. Like knowledge, the product, value, only exists there.

We can now give an intelligible account of the objective precedents. They are not flukes, but fuel. The valuing experience, like combustion, does indeed feed on natural properties and proceeds in keeping with their potential. Though emergent, it is not adventitious. The waterfall, the cardinal, the columbine, the blackberry, the warming sun, glycolysis, photosynthesis — all have indeed their stimulating properties, and thus are rightly valued when they are valued, but not until then does value appear. Perhaps, too, valuing can fail. But everything is potential until clinched in experience. Consciousness ignites what before were only combustible materials, and value lights up. The precondition for value need not itself be a value.

To say that wood is combustible means that wood will burn if ignited, although it never nears fire. This is a predicate of objective potential; wood might ignite in the spontaneous course of nature. But to say that wood is valuable is a predicate of subjective potential. If a human subject appears in relation, wood can be valued. This sort of dispositional predicate can be realized only in human experience. Some exception can here be made for subhuman experience. Animals may not have aesthetic, moral, philosophical or religious sensibilities. They may be incapable of normative discourse. But they can undergo pain and pleasure, they have interested concerns. To this extent, they own values. Valuing thus dilutes across the simplifying of the central nervous system, but, if we rely entirely on the emergent account, value is never extraneural. Where there are no centers of experience, valuing ceases and value vanishes.

But nor will this account do in explanation of the main body of natural values. While it may be true that some ranges of value emerge, like the capacity for joy or aesthetic experience, these are capstone goods, but are built on valuable substructures. There are values that only come with consciousness, but it does not follow that consciousness, when it brings its new values, confers all value and discovers none.

4. *Natural value as an entrance.* We can best appraise the emergent account in the light of another account in which value is more generously allocated to the natural world. The arriving beholder enters into the surrounding scene; it enters him. There is a two-way entrance and resulting fulfillment. Subjective experience emerges to appreciate what was before unappreciated. But such valuing is a partnership and the free-standing objective partner cannot enjoin value upon the subjective partner if it has nothing to offer. Emergence is not the whole story, there is a joining of situational value. If emergence is a *dispositional* account, we can call this a more ecological, *positional* account.

An ecologist might say that the eater realizes the potential in black-berries, but he will equally say that the eater captures nutrients instituted functionally into the ecosystem. The experienced taste is an overlay on objective food chains. The eater is waking up in the midst of events which precede and exceed his awareness. The eating of the berries, like the burning of the wood, is really a matter of formed energy throughput, a physical energy onto which life has been modulated. Initially received as solar input, nature has, by photo-synthesis, locked this energy into cellulose and carbohydrates. When humans overtake it, energy previously there is transformed in the eating and ignition. This flow-through model is a more basic one than the model of emergence. The potential is to be conceived of as a kind of capital on which we can draw a check. But the check cashing does not entirely constitute the value, reconstitute it though it may.

We ought not to forget the noblest step, but we ought not to mistake the last step for the whole history. Valuing is not apart from the whole, it is a part in the whole. Value is not isolable into a miraculous epiphenomenon or echo, even though some valued events may be happenstance. It is systemically grounded in major constructive thrusts in nature. The most satisfactory account is an ecocentric model, one which recognizes the emergence of consciousness as a novel value, and also finds this consciousness to make its entrance into a realm of objective natural value. This account works equally as well where we value things which we do not consume. When we value a thrush singing in the wild, we have a sense of entrance into events ongoing, independently of our subjective presence. We cannot genuinely care here, unless we care what happens after we are gone.

5. *Natural value as an education.* A natural object has no frame or pedestal; much depends on how I take it. The hawk flies past, and I can follow to admire its strength and speed, or let it pass and gaze into the blue expanse, pondering its fleeting smallness in the vast emptiness. Lying on my back, resting trailside, the stalwart ponderosa pine strikes me with its strength. It has stood the wintry storms. But then a hummingbird flits on scene, and how am I to interpret this interruption? By the contrast of great and small, mobile and immobile, or by comparison of different strengths? The bird has stood the winter by flight from it and arrives after five thousand miles over land and sea. I remark to my companion that this is a strong flight for so tiny a creature. But she has seen nothing. With eyes closed, she has been

wondering whether the Swainson's or the hermit thrush is the better singer.

The Fibonacci series in the spiral nebula in Andromeda can be drawn into association with that spiral in the chambered nautilus, in weather cyclones and waterfall whirlpools. I can dwell on the galaxy's size and age, on the nautilus' age and smallness, on the local whirlpool's being driven by the global Coriolis force. Natural objects trigger imaginative musings of discovery and theoretical recombination, depending upon an active following of the show, on cultural preconditioning and an adventurous openness.

Natural events thus educate us, leading out the beholder into self-expression. But it would be a mistake to conclude that all values derive entirely from our *composition* and none from our *position*. There are valued states of consciousness, but some are directed from the outside in essential, though not absolute, ways by the natural objects of consciousness. The situation remains a providing ground and catalyst, and also a check on experience. We can be be deceived, as we could not if we were only composing. If, through the floating mists at evening, I am appreciating the moon hanging over the summit, only to discover with a bit of clearing that this was the disc of a microwave antenna, I judge the experience to have been false and cannot afterwards regain it. I may be deceived about strength in the ponderosa or hummingbird. Our value judgments have to be more or less adequate to the natural facts.

Nature presents us with superposed possibilities of valuing, only some of which we realize. It is both a provocative source of and a resource for value. Here, fertility is demanded in the subject but is also found in the object which fertilizes his experience. Nature does indeed challenge us to respond as artists, poets, philosophers, and as evaluators. But rather than devaluing nature, just this educational ferment deepens its valued dimensions. The self has its options of where to take the experience nature launches, but only interactively with nature carrying forward the show. There is trail blazing by the conscious self, but also we go in the track of our surroundings, with consciousness a trailer of what lies around.

The notion that nature is a value carrier is ambiguous. Everything depends on a thing's being more or less structurally congenial for the carriage. Various items — logs, rocks, horses — will support the body and serve as a seat. Other values require rather specific carriers, for one cannot enjoy symmetry, display of colour, or adventure everywhere in nature. Still others require pregnancy with exactly that natural kind, as when only the female body can carry a child. Nature both offers and constrains values, often surprising us. We *value* a thing to discover that we are under the sway of its *valence,* inducing our behaviour. It has among its "strengths" (Latin: *valeo,* be strong) this capacity to carry value to us, if also to carry values we assign to it. This "potential" cannot always be of the empty sort which a glass has for carrying water. It is often a pregnant fullness. In the energy-throughput model, nature is indeed a carrier of value, but just as it is also objectively a carrier of energy and of life.

In climax, the values assigned to nature are up to us. But fundamentally there are powers in nature which move to us and through us. According to Alfred North Whitehead, America became great when the pioneers entered "an empty continent, peculiarly well suited for European races".[13] That suggests a vast, valueless continent, waiting to carry our imported values, although even this is belied by its being a "peculiarly well-suited" emptiness. John Locke described wild America as a "waste [where] nature and the earth furnished only the almost worthless materials as in themselves." The Europeans' labours added 999 parts of the value; hardly one part in a thousand is natural.[14] However, under our model we ought to think of a majestic and fertile ecosystem, the natural values of which could blend with those of the immigrants. Only the ecologically naive would view the energy flow, the work done, the value mix on a farm in Locke's terms. The farmer but redirects natural sources, soil fertility, sun and rain, and genetic information, to his own advantage. Even the settlers could call it (borrowing a Biblical phrase) a land of promise. Indeed, we have sometimes found values so intensely delivered that we have saved them wild, as in the Yellowstone, the Sierras and the Smokies. The cathedrals were the gems of Europe, left behind; but the national parks are the gems of America, left untouched and positively treasured for their natural value.

How shall we judge our theory that value is (in part) provided objectively in nature (T_o) against the counterbelief that value arises only as a product of subjective experience (T_s), albeit relationally with nature? Even in scientific theories, hard proof is impossible. All we can hope for is a theory from which we can logically infer certain experiences (E). If T, then E. Given these, our theory is corroborated by a kind of weak, backtracking verification. Given counterevidence (not E), we have to estimate whether the anomaly is serious. No big theory even in science, much less in value theory, is trouble free, and the theory of objective value can be stung by our seeming incapacity to know anything whatsoever in naked objectivity. But value is not the sort of thing one would expect to know without excitement. If there is objective value in nature (T_o), then one would predict it to stir up experience (E). But sometimes too that experience fails (not E), and we must presume a faulty registration and/or valueless parts of nature.

If value arrives only with consciousness (T_s), we have no problem with its absence in nature (not E). But experiences where we do find value (E), have to be dealt with as "appearances" of various sorts. The value has to be relocated in the valuing subject's creativity as he meets a valueless world, or even a valuable one; that is, one *able* to be *valued,* but which, before our bringing value ability, contains only that possibility and not any actual value. This troubles the logic by hiding too much in words such as "epiphenomenon", "echo", "emergent" and "potential". They occasionally help, but in the end give us the valuing subject in an otherwise (yet) valueless world, an insufficient premise for our experienced conclusion.

Resolute subjectivists cannot, however, be defeated by argument, although they can perhaps be driven towards analyticity. One can always hang on to the claim that value, like a tickle or remorse, must be felt to be there. Its *esse* is *percipi.* Nonsensed value is nonsense. It is impossible by argument to dislodge anyone firmly entrenched in this belief. That theirs is a retreat to definition is difficult to expose, because here they seem to cling so closely to inner experience. They are reporting, on this hand, how values always touch us. They are giving, on that hand, a stipulative definition. That is how they choose to use the word "value". At this point, discussion can go no further.

Meanwhile, the conversion to our view seems truer to world experience and more logically compelling. Here, the order of knowing reverses, if it also enhances, the order of being. This too is a perspective, but it is ecologically better informed. Nor is it so stiffly humanist and antireductionist. Science has been steadily showing how the consequents (life, mind) are built on their precedents (energy, matter), however much they overleap them. We find no reason to say that value is an irreducible emergent at the human (or upper-animal) level. We reallocate value across the whole continuum. It increases in the emergent climax but is continuously present in the composing precedents.

V
Intrinsic Natural Value

Intrinsic contrasts with *instrumental; subjective* with *objective;* and we next map these terms onto each other and the natural world.[15] *Intrinsic* value may be found in *human* experiences which are enjoyable in themselves, not needing further instrumental reference – an evening at the symphony, or one listening to loons call. Beyond this, *intrinsic natural* value recognizes value inherent in some natural occasions, without contributory human reference. The loons ought to continue to call, whether heard by humans or not. But the loon, while nonhuman, is itself a natural subject. There is something it is like to be a loon; its pains and pleasures are expressed in the call. Those who cannot conceive of nonexperienced value may allow nonhuman but not nonsubjective value. Value exists only where a subject has an object of interest.

> The being liked, or disliked of the object is its value. . . . Some sort of a subject is always requisite to there being value at all – not necessarily a *judging* subject, but a subject capable of at least motor-affective response. For the cat the cream has value, or better and more simply, the cat values the cream, or the warmth, or having her back scratched, quite regardless of her probable inability to conceive cream or to make judgments concerning warmth.[16]

Centres of experience vanish with simpler animals. In the botanical realm, we find programmes promoted, life courses generated and held to, steering cores which lock onto an individual centredness. There is a kind of "object with will", even though the feeling is gone. Every genetic set is, in that sense, a normative set; there is some "ought to be" beyond the "is", and so the plant grows, repairs itself, reproduces and defends its kind. If, after enjoying the *Trillium* in a remote woods, I step around to let it live on, I agree with this defense, and judge that here is intrinsic objective value, valued *by me,* but *for* what it is *in itself.* Value attaches to a nonsubjective form of life, but is nevertheless owned by a biological individual, a thing in itself. These things count, whether or not there is anybody to do the counting. They take account of themselves. They do what they will, which we enjoy being let in on, and which we care to see continue when we pass on. Even a crystal is an identifiable, bounded individual, a natural kind which I may wish to protect, although it has no genetic core.

However, the "for what it is in itself" facet of "intrinsic" becomes problematic in a holistic web; it is too internal and elementary; it forgets relatedness and externality. We value the humus and brooklet because in that matrix the *Trillium* springs up. They supply nutrients and water for the lake on which the loons call. With concern about populations, species, gene pools, habitats, we need a corporate sense which can also mean "good in community". Every intrinsic value has leading and trailing "ands" pointing to values from which it comes and towards which it moves. A natural fitness and positioning make individualistic intrinsic value too system independent. Neither single subject nor single object is alone. Everything is good in a role, in a whole, although we can speak of intrinsic goodness wherever a point experience, as of the *Trillium,* is so satisfying that we pronounce it good without need to enlarge our focus. Here, while experience is indeed a value, a thing can have values that go unexperienced. Just as a human life can have meaning of which the individual is unaware (for indeed the lives of all great persons have more meaning than they know), biological individuals can play valuable genetic, ecological and evolutionary roles of which they are unaware. If the truth could be known, not only is much of value taking place in nonsentient nature, much of value is going on over our own heads as well.

For comprehensive scope, let us speak of natural *projects*: some are *subjects* (loons); some are individual organic objects (*Trilliums*); some are individual material *objects* (crystals); some are *communities* (the oak-hickory forest); and some are *landforms* (Mount Rainier). Every natural affair does not have value, but there are "clots" in nature, sets of affinities with projective power, systems of thrust, counterthrust and structure to which we can attach "natures" in the plural. There are achievements with beginning, endings and cycles, more or less. Some do not have wills or interests, but rather headings, trajectories, traits and successions which give them a tectonic integrity. They are projective systems, if not selective systems. This inorganic fertility produces complexes of value — a meandering river, a string of paternoster lakes — which are reworked over time. Intrinsic value need not be immutable. Anything is of value here which has a good story to it. Anything is of value which has intense harmony, or is a project of quality. There is a negentropic constructiveness in dialectic with an entropic teardown, a mode of working for which we hardly yet have an adequate scientific, much less a valuational, theory. Yet this is nature's most striking feature, one which ultimately must be valued and of value. In one sense we say that nature is indifferent to planets, mountains, rivers and

Trilliums, but in another sense nature has bent towards making and remaking them for several billion years. These performances are worth noticing — remarkable, memorable — and they are not worth noticing just because of their tendencies to produce something else, certainly not merely because of their tendency to produce this noticing in our subjective human selves. All this gets at the root meaning of nature, its power to "generate" (Latin: *nasci, natus*).

Intrinsic natural value is a term which presides over a fading of subjective value into objective value, but also fans out from the individual to its role and matrix. Things do not have their separate natures merely in and for themselves, but they face outward and co-fit into broader natures. "Value in itself" is smeared out to become "value in togetherness". Value seeps out into the system, and we lose our capacity to identify the individual, whether subject or object, as the sole locus of value. A diagram can only suggest these diverse and complex relationships in their major zones (see figure 6). The boundaries need to be semipermeable surfaces, and there will be arrows of instrumental value (╱, ╲) found throughout, connecting occasions of individual intrinsic value (o). Each of the upper levels includes and requires much in those below it. The upper levels do not exist independently or in isolation, but only as supported and maintained by the lower levels, though the diagram, while showing this, inadequately conveys how the higher levels are perfused with the lower ones.

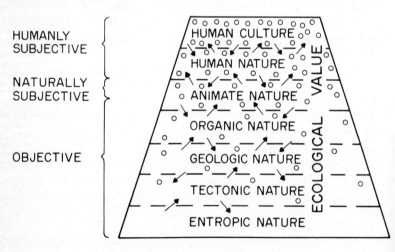

HUMANLY SUBJECTIVE

NATURALLY SUBJECTIVE

OBJECTIVE

HUMAN CULTURE
HUMAN NATURE
ANIMATE NATURE
ORGANIC NATURE
GEOLOGIC NATURE
TECTONIC NATURE
ENTROPIC NATURE

ECOLOGICAL VALUE

Figure 6. Levels of value in projective nature.

The subjectivist claim might seem safer in view of the breakout problem, that of knowing what nonexperienced value is like. But it is just as bold, for it too refuses to shut value judgments off at the boundaries of experience. It asserts a descriptive, cognitive truth about the external world of nonexperience. It too is a metaphysical claim, going beyond immediate experience to judge what is not there. Science, strictly speaking, brings a null result here; a nonanswer, not a negative answer. The subjectivist claim is certainly not simple, but rather an advanced judgment made with heavy theories replacing the primary fact of experience, where we move through a world of helps and hurts always coming at us. The logic by which one reads values out of nature is no less troublesome than the logic by which one finds values there and lets them stay.

The finding of objective value in nature is simpler and even more scientific if made with reserve. Neither native experience nor science pushes the dark back very far, but both let us in on workings which include but transcend our own existence. Immediate, middle-range experience enjoys many natural values, and one would expect this to be pragmatically competent on the local scale. When science passes to the atomic or astronomical scales, we may wish to be agnostic. But the global sciences describe an evolutionary ecosystem where, from an inchoate planet and seed of microscopic beginnings, there progressively evolves the many-splendoured panorama in which all our valuing takes place. We remain ignorant about many dynamisms and contingencies here; about what inevitability, if any, attaches to what actually did manage to happen. Nevertheless, whether rich by destiny or chance, or both, here we are, embedded in it all.

Now it does not seem simple, scientific or even safe to conceive of ourselves as subjects in metaphysical and valuational isolation from our natural launchings and underpinnings. Here, value is a powerfully penetrant notion. It slips by the emergent steps back to the generating power, away from the subject over to the web and pyramid. We certainly interpret the show, experiencing redness out of wave-lengths, beauty out of the patterned landscape. But we may be just as certain that the word known in sensory and intellectual paraphrase is structurally more complex than what comes through to register as fact and value. In that sense, in our knowing we are simplifying what is there, not enriching it, though, in another sense, the coming of humans enriches the drama, because valuers arrive in whom nature becomes conscious of itself.

In an otherwise admirable account, C. I. Lewis hedges, and grants that natural objects carry, objectively, *extrinsic* value, in effect, the standing possibility of valuation. They actually have a potential for value, even if this forever remains unexperienced or is mistakenly experienced. When an experiencer arrives, such objects do not refer us away from themselves, but we enjoy them for what they are. Nevertheless, they cannot own any intrinsic value. *"No objective existent has strictly intrinsic value; all values in objects are extrinsic only. . . . . The goodness of good objects consists in the possibility of their leading to some realization of directly experienced goodness."* Value judgments are based upon facts "obdurate and compelling" and in this sense "valuation is a form of empirical knowledge". The notion that values are only subjective is "one of the strangest aberrations ever to visit the mind of man".[17]

The word "extrinsic" suggests that, essentially still, value is a result of the human coming, whereas in ecological fact the human arrives often to trail naturally rooted values. There is nothing extraneous or accidental about the food value in a potato. When we overtake it, we recycle and amplify a natural value.[18] In evolutionary fact, there is nothing inessential or adventitious about those projective, prolife forces. They inhere in the earth itself and we latecomers inherit their work. The flow-through model of value does not find the objective side extrinsic and the subjective side intrinsic, but they are facets of one process. If, however, we revise Lewis's use of the word "extrinsic" to refer to that contributory role which natural things have, to their outward facing as this complements an inner facing, then in spontaneous nature things regularly have extrinsic as well as intrinsic value.

We can test our intuitions here by driving them to moral extremes. Let us imagine, in thought experiment, a parable of the last judgment. Suppose, a century hence, that in a tragic nuclear war each side has loosed upon the other radioactive fallout which sterilizes the genes of humans and mammals but is harmless to the flora, invertebrates, reptiles and birds. That last race of valuers, if they had conscience still, ought not to destroy the remaining biosphere. Moreover, this would not result from interest in whatever slight subjectivity might remain, for it would be better for those remaining ecosystems to continue, even if the principal valuers were taken out. That verdict would recall the Genesis parable of the first judgment in which, stage by stage, from lesser to higher forms, goodness is found at every level.

VI
The Ethic Imperative

Future historians will find our century remarkable for its breadth of knowledge and narrowness of value judgments. Never have humans known so much about, and valued so little in, the great chain of being. As a result, the ecological crisis is not surprising. To devalue nature and inflate the human worth is to do business in a false currency. This yields a dysfunctional, monopolistic world view. We are misfits because we have misread our life-support system. We rationalize that the place we inhabit has no normative structures, and that we can do what we please. Afterwards, this view sinks down into the hinterlands of our minds, an invisible persuader which silently shapes an ethic.

One can blunder in the old, naive view that values are known in literal, uninterpreted simplicity. But there is folly also in swinging to the other extreme. In this arrogation of value to ourselves, there is what the theologians call *hubris,* overbearing pride. It is much easier to impose our wills on the world when we believe it is otherwise of no account. Nothing stays our libido; nothing demands any human-transcending concern. But ethics too, like all aspects of life, flourishes when operating in a system of checks and balances. What if, in truth, we are not only limited by the natural facts but also by natural values? What if living well is not merely a getting of what is valued, but a negotiating of values in a neighbourhood of worth? In the former belief we would forever remain juvenile. In the latter we should gain moral maturity.

There is much nobility in being self-actualizing, and nature permits us to elect some values. But such dignity is not enhanced by living as lonesome selves in a void world. There is no joy in being freaks of nature, loci of value lost in a worthless environment. The doctrine of the sterility of nature is not a boon but an evil, for it throws humans into meaninglessness, into an identity crisis. It has made much in modern life sterile. At this point, there is something encouraging about the notion of relativity. Einstein introduced us, at the levels of time, space, mass and energy, to but one form of an ecosystemic principle. Subjectivity, too, is what it is in objective circumstances. The values we own are nested in a mother matrix. To turn Bishop Butler on his head: everything is what it is in relation to other things.[19] This kind of relativity does not cause alienation and anthropocentrism; it rather cures it.

Seen in this way, it is not the objectivists but rather the thorough-going subjectivists who uphold the naturalistic fallacy. They must either derive value at a consummate stroke out of a merely factual nature, getting it, as it were, *ex nihilo,* or out of something available but to no avail without us; or they have to bring value in by skyhook from some *a priori* source. But we do not commit this fallacy because we find fact and value inseparably to coevolve. This does not deny the mystery of emerging value, but there is value in our premises as well as in our conclusion.

We humans do not play out our drama of epiphenomenal or emergent value on a valueless natural stage. The stage is the womb from whence we come, but which we really never leave. If the enduring drama has any value, that must somehow attach to the whole plot and plasma, span over from potential to persons, even though it may be diversely distributed across events. Nature is not barren of value; it is rather the bearer of value: that both constrains and ennobles the role we humans are called to play.

Notes

1. William James, *The Varieties of Religious Experience* (New York: Longmans, Green and Co., 1925), p. 150.
2. John Laird, *A Study in Realism* (Cambridge: Cambridge University Press, 1920), p. 129.
3. Wilhelm Windelband, *An Introduction to Philosophy,* trans. Joseph McCabe (London: T. Fisher Unwin Ltd., 1921), p. 215.
4. Ralph Barton Perry, *General Theory of Value* (Cambridge, Mass.: Harvard University Press, 1926), pp. 125, 115–16.
5. Samuel Alexander, *Beauty and Other Forms of Value* (New York: Thomas Y. Crowwell Company, 1968), pp. 172–87.
6. Cited in Ernst Cassirer, *Substance and Function and Einstein's Theory of Relativity* (New York: Dover Publications, 1953), p. 356.
7. John Wheeler, "From Relativity to Mutability", in *The Physicist's Conception of Nature,* ed. Jagdish Mehra (Dordrecht, Holland: D. Reidel Publishing Co., 1973), p. 227.
8. Werner Heisenberg, "The Representation of Nature in Contemporary Physics", *Daedalus* 87 no. 3 (Summer 1958): 107.
9. Lubert Stryer, *Biochemistry* (San Francisco: W. H. Freeman and Co., 1975), p. 90.
10. John Dewey, *Experience and Nature* (New York: Dover Publications, 1958), p. 4a.
11. This seems to be the view of George Santayana, *The Sense of Beauty* (New York: Modern Library, 1955), pp. 21–24, 150–54.
12. Alexander, *Beauty,* pp. 30–31. Alexander holds, however, that certain non-

aesthetic values may exist in nature, pp. 285–99; see also, idem, *Space, Time, and Deity*, 3 vols. (London: Macmillan and Co., 1920), 2: 302–14.

13. Alfred North Whitehead, "The Study of the Past – Its Uses and Its Dangers", *Harvard Business Review* 11 (1932–33): 438.

14. John Locke, *The Second Treatise of Civil Government* (Oxford: Basil Blackwell, 1948), secs. 42, 43, pp. 22–23.

15. We set aside a use of *subjective* which means "depending on personal judgment, difficult to get consensus on". By contrast, *objective* means "obvious to all, publicly demonstrable". Many instrumental, humanistic values in nature – our need for food – are unarguable, while finding intrinsic natural value requires discretionary, subjective judgment. We here examine subjectivity in claim content, not that involved in verifying a claim. It is not surprising that humans reach the easiest consensus on values nearest those we subjects experience, nor that there is disagreement about objective value, since nonexperienced value is remote from the immediacy of personal life.

16. David W. Prall, *A Study in the Theory of Value* (Berkeley and Los Angeles: University of California Press, 1921), p. 227.

17. Clarence Irving Lewis, *An Analysis of Knowledge and Valuation* (La Salle, Illinois: Open Court Publishing Co., 1946), p. 387; ibid, p. 407; ibid, p. vii; ibid, p. 366.

18. The apple, Lewis would reply, cannot realize its own value. If uneaten, it rots. So its value is extrinsic; the eater's pleasure is intrinsic. But this example is misleading unless ecologically understood. The carbohydrate stored in the overwintering potato will be used, although not experienced, by the plant in the spring. Eating it overtakes energy of value to the plant. But the apple functions as a gamble in seed dispersal. Its value is realized when birds, deer or humans take the bait. The apple has been very successful; it has caught the man. While the apple takes care of the man, the man takes care of the apple. Its survival is assured as long as there are humans to care for it!

19. "Every thing is what it is, and not another thing" (Joseph Butler, preface to *Fifteen Sermons upon Human Nature* [London, 1726]).

Duties Concerning Islands
MARY MIDGLEY

Had Robinson Crusoe any duties?

When I was a philosophy student, this used to be a familiar conundrum, which was supposed to pose a very simple question; namely, can you have duties to yourself? Mill, they correctly told us, said no. "The term duty to oneself, when it means anything more than prudence, means self-respect or self-development, and for none of these is anyone accountable to his fellow-creatures."[1] Kant, on the other hand, said yes. "Duties to ourselves are of primary importance and should have pride of place . . . nothing can be expected of a man who dishonours his own person."[2] There is a serious disagreement here, not to be sneezed away just by saying, "it depends on what you mean by duty". Much bigger issues are involved — quite how big has, I think, not yet been fully realized. To grasp this, I suggest that we rewrite a part of Crusoe's story, in order to bring in sight a different range of concerns.

> 19 Sept. 1685. This day I set aside to devastate my island. My pinnace being now ready on the shore, and all things prepared for my departure, Friday's people also expecting me, and the wind blowing fresh away from my little harbour, I had a mind to see how all would burn. So then, setting sparks and powder craftily among certain dry spinneys which I had chosen, I soon had it ablaze, nor was there left, by the next dawn, any green stick among the ruins. . . .

Now, work on the style how you will, you cannot make that into a convincing paragraph. Crusoe was not the most scrupulous of men, but he would have felt an invincible objection to this senseless destruction. So would the rest of us. Yet the language of our moral tradition has tended strongly, ever since the Enlightenment, to make that objection unstateable. All the terms which express that an obligation is serious or binding — duty, right, law, morality, obligation, justice — have been deliberately narrowed in their use so as to apply only in the framework of contract, to describe only relations holding between free and rational agents. Since it has been decided *a priori* that rationality admits of no degrees and that cetaceans are not rational, it follows that, unless you take either religion or science fiction seriously, we can only have duties to humans, and sane, adult, responsible humans at that. Now the morality we live by certainly does not accept this restriction. In common life we recognize many other duties as serious and binding, though of course not necessarily overriding. If philosophers want to call these something else instead of duties, they must justify their move.

We have here one of those clashes between the language of common morality (which is of course always to some extent confused and inarticulate) and an intellectual scheme which arose in the first place from a part of that morality, but has now taken off on its own claims of authority to correct other parts of its source. There are always real difficulties here. As ordinary citizens, we have to guard against dismissing such intellectual schemes too casually; we have to do justice to the point of them. But, as philosophers, we have to resist the opposite temptation of taking the intellectual scheme as decisive, just because it is elegant and satisfying, or because the moral insight which is its starting point is specially familiar to us. Today, this intellectualist bias is often expressed by calling the insights of common morality mere "intuitions". This is quite misleading, since it gives the impression that they have been reached without thought, and that there is, by contrast, a scientific solution somewhere else to which they ought to bow — as there might be if we were contrasting commonsense "intuitions" about the physical world with physics or astronomy. Even without that word, philosophers often manage to give the impression that whenever our moral views clash with any simple, convenient scheme, it is our *duty* to abandon them. Thus, Grice states:

> It is an inescapable consequence of the thesis presented in these pages that certain classes cannot have natural rights: animals, the human embryo, future generations, lunatics and children under the age of, say, ten. In the case of young children at least, my experience is that this consequence is found hard to accept. But it is a consequence of the theory; it is, I believe, true; and I think we should be willing to accept it. At first sight it seems a harsh conclusion, but it is not nearly so harsh as it appears.[3]

But it is in fact extremely harsh, since what he is saying is that the treatment of children ought not to be determined by their interests but by the interests of the surrounding adults capable of contract, which, of course, can easily conflict with them. In our society, he explains, this does not actually make much difference, because parents here are so benevolent that they positively want to benefit their children, and accordingly here "the interests of children are reflected in the interests of their parents". But this, he adds, is just a contingent fact about us. "It is easy to imagine a society where this is not so", where, that is, parents are entirely exploitative. "In this circumstance, the morally correct treatment of children would no doubt be harsher than it is in our society. But the conclusion has to be accepted." Grice demands that we withdraw our objections to harshness, in deference to theoretical consistency. But "harsh" here does not mean just "brisk and bracing", like cold baths and a plain diet. (There might well be more of those where parents do feel bound to consider their children's interests.) It means "unjust". Our objection to unbridled parental selfishness is not a mere matter of tone or taste; it is a moral one. It therefore requires a moral answer, an explanation of the contrary *value* which the contrary theory expresses. Grice, and those who argue like him, take the ascetic, disapproving tone of those who have already displayed such a value, and who are met by a slovenly reluctance to rise to it. But they have not displayed that value. The ascetic tone cannot be justified merely by an appeal to consistency. An ethical theory which, when consistently followed through, has iniquitous consequences, is a bad theory and must be changed. Certainly we can ask whether these consequences really are iniquitous, but this question must be handled seriously. We cannot directly conclude that the consequences cease to stink the moment they are seen to follow from our theory.

The theoretical model which has spread blight in this area is, of course, that of social contract, and, to suit it, that whole cluster of essential moral terms − right, duty, justice and the rest − has been progressively narrowed. This model shows human society as a spread of standard social atoms, originally distinct and independent, each of which combines with others only at its own choice and in its own private interest. This model is drawn from physics, and from seventeenth-century physics, at that, where the ultimate particles of matter were conceived as hard, impenetrable, homogeneous little billiard balls, with no hooks or internal structure. To see how such atoms could combine at all was very hard. Physics, accordingly, moved on from this notion to one which treats atoms and other particles as complex items, describable mainly in terms of forces, and those the same kind of forces which operate outside them. It has abandoned the notion of ultimate, solitary, independent individuals. Social-contract theory, however, retains it.

On this physical – or archaeo-physical – model, all significant moral relations between individuals are the symmetrical ones expressed by contract. If, on the other hand, we use a biological or "organic" model, we can talk also of a variety of asymmetrical relations found within a whole. Leaves relate not only to other leaves, but to fruit, twigs, branches and the whole tree. People appear not only as individuals, but as members of their groups, families, tribes, species, ecosystems and biosphere, and have moral relations as parts to these wholes. The choice between these two ways of thinking is not, of course, a simple once-and-for-all affair. Different models are useful for different purposes. We can, however, reasonably point out, firstly, that the old physical pattern does make all attempts to explain combination extremely difficult; and, secondly, that since human beings actually are living creatures, not crystals or galaxies, it is reasonable to expect that biological ways of thinking will be useful in understanding them.

In its own sphere, the social contract model has of course been of enormous value. Where we deal with clashes of interest between free and rational agents already in existence, and particularly where we want to disentangle some of them from some larger group that really does not suit them, it is indispensable. And for certain political purposes during the last three centuries these clashes have been vitally important. An obsession with contractual thinking, and a conviction that it is a cure-all, are therefore understandable. But the trouble with such obsessions is that they distort the whole shape of thought and language in a way which makes them self-perpetuating, and constantly extends their empire. Terms come to be defined in a way which leaves only certain moral views expressible. This can happen without any clear intention on the part of those propagating them, and even contrary to their occasional declarations, simply from mental inertia. Thus, John Rawls, having devoted most of his long book to his very subtle and exhaustive contractual view of justice, remarks without any special emphasis near the end that "we should recall here the limits of a theory of justice. Not only are many aspects of morality left aside, but no account can be given of right conduct in regard to animals and the rest of nature."[4] He concedes that these are serious matters. "Certainly it is wrong to be cruel to animals and the destruction of a whole species can be a great evil. The capacity for feelings of pleasure and pain and for the forms of life of which animals are capable clearly impose duties of compassion and humanity in their case." All this is important, he says, and it calls for a wider metaphysical enquiry, but it is not his subject. Earlier in the same passage he touches on the question of permanently irrational human beings, and remarks that it "may present a difficulty. I cannot examine this problem here, but I assume that the account of equality would not be materially affected."[5] Won't it though? It is a strange project to examine a single virtue — justice — without at least sketching in one's view of the vast background of general morality which determines its shape and meaning, including, of course, such awkward and noncontractual virtues as "compassion and humanity". It isolates the duties which people owe each other *merely as thinkers* from those deeper and more general ones which they owe each other as beings who feel. It cannot, therefore, fail both to split a man's nature and to isolate him from the rest of the creation to which he belongs.

Such an account may not be *Hamlet* without the prince, but it is *Hamlet* with half the cast missing, and without the state of Denmark. More exactly, it is like a history of Poland which regards Russia, Germany, Europe and the Roman Church as not part of its subject. I am not attacking Rawls' account on its own ground. I am simply pointing out what the history of ethics shows all too clearly — how much our thinking is shaped by what our sages *omit* to mention. The Greek philosophers never really raised the problem of slavery till towards the end of their speech, and then few of them did so with conviction. This happened even though it lay right in the path of their enquiries into political justice and the value of the individual soul. Christianity did raise that problem, because its class background was different and because the world in the Christian era was already in turmoil, so that men were not presented with the narcotic of a happy stability. But Christianity itself did not, until quite recently, raise the problem of the morality of punishment, and particularly of eternal punishment. This failure to raise central questions was not, in either case, complete. One can find very intelligent and penetrating criticisms of slavery occurring from time to time in Greek writings — even in Aristotle's defence of that institution.[6] But they are mostly like Rawls' remark here. They conclude that "this should be investigated some day". The same thing happens with Christian writings concerning punishment, except that the consideration, "this is a great mystery", acts as an even more powerful paralytic to thought. Not much more powerful, however. Natural inertia, when it coincides with vested interest or the illusion of vested interest, is as strong as gravitation.

It is important that Rawls does not, like Grice, demand that we toe a line which would make certain important moral views impossible. Like Hume, who similarly excluded animals from justice, he simply leaves them out of his discussion. This move ought in principle to be harmless. But when it is combined with an intense concentration of discussion on contractual justice, and a corresponding neglect of compassion and humanity, it inevitably suggests that the excluded problems are relatively unimportant. This suggestion is still more strongly conveyed by rulings which exclude the nonhuman world from rights, duties and morality. Words like "rights" and "duties" are awkward because they do indeed have narrow senses approximating to the legal, but they also have much wider ones in which they cover the whole moral sphere. To say "they do not have rights", or "you do not have duties to them" conveys to any ordinary hearer a very simple message; namely, "they do not matter". This is an absolution, a removal of blame for ill-treatment of "them", whoever they may be.

To see how strong this informal, moral usage of "rights" is, we need only look at the history of that powerful notion, the "rights of man". These rights were not supposed to be ones conferred by law, since the whole point of appealing to them was to change laws so as to embody them. They were vague, but vast. They did not arise, as rights are often said to do, only within a community, since they were taken to apply in principle everywhere. The immense, and on the whole coherent, use which has been made of this idea by reform movements shows plainly that the tension between the formal and the informal idea of "right" is part of the word's meaning, a fruitful connection of thought, not just a mistake. It is therefore hard to adopt effectively the compromise which some philosophers now favour, of saying that it is indeed wrong to treat animals in certain ways, but that we have no duties to them or that they have no rights.[7] "Animal rights" may be hard to formulate, as indeed are the rights of humans. But "no rights" will not do. The word may need to be dropped entirely. The compromise is still harder with the word "duty", which is rather more informal, and is more closely wedded to a private rather than political use.

Where the realm of right and duty stops, there, to ordinary thinking, begins the realm of the optional. What is not a duty may be a matter of taste, style or feeling, of aesthetic sensibility, of habit and nostalgia, of etiquette and local custom, but it cannot be something which demands our attention whether we like it or not. When claims get into this area, they can scarcely be taken seriously. This becomes clear when Kant tries to straddle the border. He says that we have no direct duties to animals, because they are not rational, but that we should treat them properly all the same because of "indirect" duties which are really duties to our own humanity.[8] This means that ill-treating them (a) might lead us to ill-treat humans, and (b) is a sign of a bad or inhumane disposition. The whole issue thus becomes a contingent one of spiritual style or training, like contemplative exercises, intellectual practice or, indeed, refined manners.[9] Some might need practice of this kind to make them kind to people, others might not, and, indeed, might get on better without it. (Working off one's ill-temper on animals might make one treat people *better*.) But the question of cruelty to animals cannot be like this, because it is of the essence to such training exercises that they are internal. Anything that affects some other being is not just practice, it is real action. Anyone who refrained from cruelty *merely* from a wish not to sully his own character, without any direct consideration for the possible victims, would be frivolous and narcissistic.

A similar trivialization follows where theorists admit duties of compassion and humanity to noncontractors, but deny duties of justice. Hume and Rawls, in making this move, do not explicitly subordinate these other duties, or say that they are less binding. But because they make the contract element so central to morality, this effect appears to follow. The priority of justice is expressed in such everyday proverbs as "be just before you're generous". We are therefore rather easily persuaded to think that compassion, humanity and so forth are perhaps emotional luxuries, to be indulged only after all debts are paid. A moment's thought will show that this is wrong. Someone who receives simultaneously a request to pay a debt and another to comfort somebody bereaved or on their death bed is not as a matter of course under obligation to treat the debt as the more urgent. He has to look at circumstances on both sides, but in general we should probably expect the other duties to have priority. This is still more true if, on his way to pay the debt, he encounters a stranger in real straits, drowning or lying on the road. To give the debt priority, we probably need to think of his creditor as also being in serious trouble — which brings compassion and humanity in on both sides of the case.

What makes it so hard to give justice a different clientele from the other virtues, as Hume and Rawls do, is simply the fact that justice is such a pervading virtue. In general, all serious cases of cruelty, meanness, inhumanity and the like are also cases of injustice. If we are told that a certain set of these cases does not involve injustice, our natural thought is that these cases must be *trivial.* Officially, Hume's and Rawls' restriction is not supposed to mean this. What, however, is it supposed to mean? It is forty years since I first read Hume's text, and I find his thought as obscure now as I did then. I well remember double-taking then, going back over the paragraph for a point which, I took it, I must have missed. Can anyone see it?

> Were there a species of creatures intermingled with men, which, though rational, were possessed of such inferior strength, both of body and mind, that they were incapable of all resistance, and could never, upon the highest provocation, make us feel the effects of their resentment; the necessary consequence, I think, is that we should be bound by the laws of humanity to give gentle usage to these creatures, but should not, properly speaking, lie under any restraint of justice with regard to them, nor could they possess any right or property, exclusive of such arbitrary lords. Our intercourse with them could not be called society, which supposes a degree of equality, but absolute command on one side and servile obedience on the other. . . . This is plainly the situation of men with regard to animals.[10]

I still think that the word "justice", so defined, has lost its normal meaning. In ordinary life we think that duties of justice become *more* pressing, not less so, when we are dealing with the weak and inarticulate, who cannot argue back. It is the boundaries of prudence which depend on power, not those of justice. Historically, Hume's position becomes more understandable when one sees its place in the development of social-contract thinking. The doubtful credit for confining justice to the human species seems to belong to Grotius, who finally managed to ditch the Roman notion of *jus naturale,* natural right or law, common to all species. I cannot here discuss his remarkably unimpressive arguments for this.[11] The point I want to make here is simply in reference to the effect of these restrictive definitions of terms like "justice" on people's view of the sheer size of the problems raised by what falls outside them.

Writers who treat morality as primarily contractual tend to discuss noncontractual cases briefly, casually and parenthetically, as though they were rather rare. Rawls' comments on the problem of mental defectives are entirely typical here. We have succeeded, they say, in laying most of the carpet; why are you making this fuss about those little wrinkles behind the sofa? This treatment confirms a view, already suggested by certain aspects of current politics in the United States, that those who fail to clock in as normal rational agents and make their contracts are just occasional exceptions, constituting one more "minority" group — worrying, no doubt, to the scrupulous, but not a central concern of any society. Let us, then, glance briefly at their scope, by roughly listing some cases which seem to involve us in non-contractual duties. (The order is purely provisional and the numbers are added just for convenience.)

Human sector	1. The dead
	2. Posterity
	3. Children
	4. The senile
	5. The temporarily insane
	6. The permanently insane
	7. Defectives, ranging down to "human vegetables"
	8. Embryos, human and otherwise
Animal sector	9. Sentient animals
	10. Nonsentient animals
Inanimate sector	11. Plants of all kinds
	12. Artefacts, including works of art
	13. Inanimate but structured objects — crystals, rivers, rocks, etc.
Comprehensive	14. Unchosen groups of all kinds, including families and species
	15. Ecosystems, landscapes, villages, warrens, cities, etc.
	16. Countries
	17. The Biosphere
Miscellaneous	18. Oneself
	19. God

No doubt I have missed a few, but that will do to go on with. The point is this; if we look only at a few of these groupings, and without giving them full attention, it is easy to think that we can include one or two as honorary contracting members by a slight stretch of our conceptual scheme, and find arguments for excluding the others from serious concern entirely. But if we keep our eye on the size of the range, this stops being plausible. As far as sheer numbers go, this is no minority of the beings with whom we have to deal. We are a small minority of them. As far as importance goes, it is certainly possible to argue that some of these sorts of beings should concern us more and others less: we need a priority system. But, to build it, *moral* arguments are required. The various kinds of claims have to be understood and compared, not written off in advance. We cannot rule that those who, in our own and other cultures, suppose that there is a direct objection to injuring or destroying some of them, are always just confused, and mean only, in fact, that this item will be needed for rational human consumption.[12]

The blank antithesis which Kant made between rational persons (having value) and mere things (having none) will not serve us to map out this vast continuum. And the idea that, starting at some given point on this list, we have a general licence for destruction, is itself a moral view which would have to be justified. Western culture differs from most others in the breadth of destructive licence which it allows itself, and, since the seventeenth century, that licence has been greatly extended. Scuples about rapine have been continually dismissed as irrational, but it is not always clear with what rational principles they are supposed to conflict. Western destructiveness has not in fact developed in response to a new set of disinterested intellectual principles demonstrating the need for more people and less redwoods, but mainly as a by-product of greed and increasing commercial confidence. Humanistic hostility to superstition has played some part in the process, because respect for the nonhuman items on our list is often taken to be religious. It does not have to be. Many scientists who are card-carrying atheists can still see the point of preserving the biosphere. So can the rest of us, religious or otherwise. It is the whole of which we are parts, and its other parts concern us for that reason.

But the language of rights is rather ill-suited to expressing this, because it has been developed mainly for the protection of people who, though perhaps oppressed, are in principle articulate. This makes it quite reasonable for theorists to say that rights belong only to those who understand them and can claim them. When confronted with the "human sector" of our list, these theorists can either dig themselves in, like Grice, and exclude the lot, or stretch the scheme, like Rawls, by including the hypothetical rational choices which these honorary members *would* make if they were not unfortunately prevented. Since many of these people seem less rational than many animals, zoophiles have, then, a good case for calling this second device arbitrary or specious, and extending rights to the border of sentience. Here, however, the meaning of the term "rights" does become thin, and when we reach the inanimate area, usage will scarcely cover it. (It is worth noticing that long before this, when dealing merely with the "rights of man", the term often seems obscure, because to list and specify these rights is so much harder than to shout for them. The word is probably of more use as a slogan, indicating a general direction, than as a detailed conceptual tool.) There may be a point in campaigning to extend usage. But to me it seems wiser on the whole not to waste energy on this verbal point, but instead to insist on the immense variety of kinds of beings with which we have to deal. Once we grasp this, we ought not to be surprised that we are involved in many different kinds of claim or duty. The dictum that "rights and duties are correlative" is misleading, because the two words keep rather different company, and one may be narrowed without affecting the other.

What, then, about duties? I believe that this term can properly be used over the whole range. We have quite simply got many kinds of duties to animals,[13] to plants and to the biosphere. But to speak in this way we must free the term once and for all from its restrictive contractual use, or irrelevant doubts will still haunt us. If we cannot do this, we shall have to exclude the word "duty", along with "rights" from all detailed discussion, using wider words like "wrong", "right" and "ought" instead. This gymnastic would be possible but inconvenient. The issue about duty becomes clear as soon as we look at the controversy from which I started, between Kant's and Mill's views on duties to oneself. What do we think about this? Are there duties of integrity, autonomy, self-knowledge, self-respect? It seems that there are. Mill is right, of course, to point out that they are not duties *to* someone in the ordinary sense. The divided self is a metaphor. It is as natural and necessary a metaphor here as it is over, say, self-deception or self-control, but it certainly is not literal truth. The form of the requirement is different. Rights, for instance, certainly do not seem to come in here as they often would with duties to other persons; we would scarcely say, "I have a right to my own respect." And the *kind* of things which we can owe ourselves are distinctive. It is not just chance who they are owed to. You cannot owe it to somebody else, as you can to yourself, to force him to act freely or with integrity. He owes that to himself, the rest of us can only remove outside difficulties. As Kant justly said, our business is to promote our own perfection and the happiness of others; the perfection of others is an aim which belongs to them.[14] Respect, indeed, we owe both to ourselves and to others, but Kant may well be right to say that self-respect is really a different and deeper requirement, something without which all outward duties would become meaningless. (This may explain the paralyzing effect of depression.)

Duties to oneself, in fact, are duties with a different *form*. They are far less close than outward duties to the literal model of debt, especially monetary debt. Money is a thing which can be owed in principle to anybody, it is the same whoever you owe it to, and if by chance you come to owe it to yourself, the debt vanishes. Not many of our duties are really of this impersonal kind; the attempt to commute other sorts of duties into money is a notorious form of evasion. Utilitarianism however wants to make all duties as homogeneous as possible. And that is the point of Mill's position. He views all our self-concerning motives as parts of the desire for happiness. Therefore he places all duty, indeed all morality, on the outside world, as socially required restriction of that desire — an expression, that is, of other people's desire for happiness.

> We do not call anything wrong, unless we mean that a person ought to be punished in some way or another for doing it; if not by law, by the opinion of his fellow-creatures; if not by opinion, by the reproaches of his own conscience. This seems the real turning point of the distinction between morality and simple expediency. It is a part of the notion of Duty in every one of its forms, that a person may rightly be compelled to fulfil it. Duty is a thing which may be *exacted* from a person, as one exacts a debt.[15]

To make the notion of wrongness depend on punishment and public opinion in this way instead of the other way round is a bold step. Mill did not mind falling flat on his face from time to time in trying out a new notion for the public good. He did it for us, and we should, I think, take proper advantage of his generosity, and accept the impossibility which he demonstrates. The concepts cannot be connected this way round. Unless you think of certain acts as wrong, it makes no sense to talk of punishment. "Punishing" alcoholics with aversion therapy or experimental rats with electric shocks is not really punishing at all; it is just deterrence. This "punishment" will not make their previous actions wrong, nor has it anything to do with morality. The real point of morality returns to Mill's scheme in the Trojan horse of "the reproaches of his own conscience". Why do *they* matter? Unless the conscience is talking sense — that is, on Utilitarian principles, unless it is delivering the judgment of society — it should surely be silenced. Mill, himself a man of enormous integrity, deeply concerned about autonomy, would never have agreed to silence it. But, unless we do so, we shall have to complicate his scheme. It may well be true that, in the last resort and at the deepest level, conscience and the desire for happiness converge. But in ordinary life and at the everyday level they can diverge amazingly. We do want to be honest but we do not want to be put out. What we know we ought to do is often most unwelcome to us, which is why we call it duty. And whole sections of that duty do not concern other people directly at all. A good example is the situation in Huxley's *Brave New World,* where a few dissident citizens have grasped the possibility of a fuller and freer life. Nobody else wants this. Happiness is already assured. The primary duty of change here seems to be that of each to himself. True, they may feel bound also to help others to change, but hardly in a way which those others would *exact.* In fact, we may do better here by dropping the awkward second party altogether and saying that they have a duty *of* living differently — one which will affect both themselves and others, but which does not require, as a debt does, a named person or people *to* whom it must be paid. Wider models like "the whole duty of man" may be more relevant.

This one example from my list will, I hope, be enough to explain the point. I cannot go through all of them, nor ought it to be necessary. Duties need *not* be quasi-contractual relations between symmetrical pairs of rational human agents. There are all kind of other obligations holding between asymmetrical pairs, or involving, as in this case, no outside beings at all. To speak of duties *to* things in the inanimate and comprehensive sectors of my list is not necessarily to personify them superstitiously, or to indulge in chatter about the "secret life of plants".[16] It expresses merely that there are suitable and unsuitable ways of behaving in given situations. People have duties *as* farmers, parents, consumers, forest dwellers, colonists, species members, ship-wrecked mariners, tourists, potential ancestors and actual descendants, etc. As such, it is the business of each not to forget his transitory and dependent position, the rich gifts which he has received, and the tiny part he plays in a vast, irreplaceable and fragile whole.

It is remarkable that we now have to state this obvious truth as if it were new, and invent the word "ecological" to describe a whole vast class of duties. Most peoples are used to the idea. In stating it, and getting back into the centre of our moral stage, we meet various difficulties, of which the most insidious is possibly the temptation to feed this issue as fuel to long-standing controversies about religion. Is

concern for the nonhuman aspects of our biosphere necessarily super-
stitious and therefore to be resisted tooth and nail? I have pointed out
that it need not be religious. Certified rejectors of all known religions
can share it. No doubt, however, there is a wider sense in which any
deep and impersonal concern can be called religious — one in which
Marxism is a religion. No doubt, too, all such deep concerns have their
dangers, but certainly the complete absence of them has worse dangers.
Moreover, anyone wishing above all to avoid the religious dimension
should consider that the intense individualism which has focused our
attention exclusively on the social-contract model is itself thoroughly
mystical. It has glorified the individual human soul as an object having
infinite and transcendent value; has hailed it as the only real creator;
and bestowed on it much of the panoply of God. Nietzsche, who was
responsible for much of this new theology,[17] took over from the old
theology (which he plundered extensively) the assumption that all the
rest of creation mattered only as a frame for humankind. This is not an
impression which any disinterested observer would get from looking
round at it, nor do we need it in order to take our destiny sufficiently
seriously.

Crusoe then, I conclude, did have duties concerning his island, and
with the caution just given we can reasonably call them duties *to* it.
They were not very exacting, and were mostly negative. They differed,
of course, from those which a long-standing inhabitant of a country
has. Here the language of *fatherland* and *motherland,* which is so widely
employed, indicates rightly a duty of care and responsibility which can
go very deep, and which long-settled people commonly feel strongly.
To insist that it is really only a duty to the exploiting human beings is
not consistent with the emphasis often given to reverence for the actual
trees, mountains, lakes, rivers and the like which are found there. A
decision to inhibit all this rich area of human love is a special manoeuvre
for which reasons would need to be given, not a dispassionate analysis
of existing duties and feelings. What happens, however, when you are
shipwrecked on an entirely strange island? As the history of coloniza-
tion shows, there is a tendency for people so placed to drop any
reverence and become more exploitative. But it is not irresistible.
Raiders who settle down can quite soon begin to feel at home, as the
Vikings did in East Anglia, and can, after a while, become as possessive,
proud and protective towards their new land as the old inhabitants.
Crusoe himself does, from time to time, show this pride rather
touchingly, and it would, I think, certainly have inhibited any moderate

temptation, such as that which I mentioned, to have a good bonfire. What keeps him sane through his stay is in fact his duty to God. If that had been absent, I should rather suppose that sanity would depend on a stronger and more positive attachment to the island itself and its creatures. It is interesting, however, that Crusoe's story played its part in developing that same icy individualism which has gone so far towards making both sorts of attachment seem corrupt or impossible. Rousseau delighted in *Robinson Crusoe,* and praised it as the only book fit to be given to a child, *not* because it showed a man in his true relation to animal and vegetable life, but because it was the bible of individualism. "The surest way to raise him [the child] above prejudice and to base his judgments on the true relations of things, is to put him in the place of a solitary man, and to judge all things as they would be judged by such a man in relation to their own utility. . . . So long as only bodily needs are recognized, man is self-sufficing . . . the child knows no other happiness but food and freedom."[18] That false atomic notion of human psychology — a prejudice above which nobody ever raised Rousseau — is the flaw in all social-contract thinking. If he were right, every member of the human race would need a separate island — and what, then, would our ecological problems be? Perhaps, after all, we had better count our blessings.

Notes

1. John Stuart Mill, *Essay on Liberty* (London: Dent, Everyman's Library, 1910), chap. 4, p. 135.
2. Immanuel Kant, "Duties to Oneself", in *Lectures on Ethics,* trans. Louis Infield (London: Methuen, 1930), p. 118.
3. G. R. Grice, *Grounds for Moral Sentiments* (Cambridge: Cambridge University Press, 1967), pp. 147–49.
4. John Rawls, *A Theory of Justice* (Oxford: Oxford University Press, 1972), p. 512.
5. Ibid., p. 510.
6. Aristotle *Politics* 1. 3–8; cf., idem, *Nichomachian Ethics* 7, 2.
7. For example, John Passmore, *Man's Responsibility for Nature* (London: Duckworth, 1974), pp. 116–17; H. J. McCloskey, "Rights", *Philosophical Quarterly* 15 (1965).
8. Nor will it help for philosophers to say "it is not the case that they have rights". Such pompous locutions have either no meaning at all, or the obvious one.
9. Immanual Kant, "Duties towards Animals and Spirits", in *Lectures on Ethics,* p. 240.

10. David Hume "An Enquiry Concerning the Principles of Morals, in *Hume's Moral and Political Philosophy,* ed. H.E. Aiben (New York: Hafner, 1949), app. 3, pp. 190–91.

11. A point well discussed by Stephen R. L. Clark, *The Moral Status of Animals* (Oxford: Clarendon Press, 1977), pp. 12–13.

12. For details, see John Rodman, "Animal Justice: The Counter-Revolution in Natural Right and Law", *Inquiry* 22, nos. 1–2 (Summer 1979).

13. A case first made by Jeremy Bentham, *An Introduction to the Principles of Morals and Legislation,* Chap. 17, and well worked out by Peter Singer, *Animal Liberation* (New York: Avon, 1975), Chaps. 1, 5 and 6.

14. Immanuel Kant, *Preface to the Metaphysical Elements of Ethics,* sec. "Introduction to Ethics", 4 and 5.

15. John Stuart Mill, *Utilitarianism* (London: Dent, Everyman's Library, 1910), chap. 5, p. 45.

16. P. Tompkins and C. Bird, *The Secret Life of Plants* (New York: Harper and Row, 1973), claimed to show, by various experiments involving electrical apparatus, that plants can feel. Attempts to duplicate their experiments have, however, totally failed to produce any similar results. (See A. W. Galston and C. L. Slayman, "The Secret Life of Plants", *American Scientist* 67 [1973]: 337). It seems possible that the original results were due to a fault in the electrical apparatus. The attempt shows, I think, one of the confusions which continually arise from insisting that all duties must be of the same form. We do not need to prove that plants are animals in order to have reason to spare them. The point is well discussed by Marian Dawkins in her book *Animal Suffering* (London: Chapman and Hall, 1981), pp. 117–19.

17. See particularly, Friedrich Nietzsche *Thus Spake Zarathurstra* 3, section "Of Old and New Tables"; and *The Joyful Wisdom* (Otherwise called *The Gay Science*), p. 125 (the Madman's Speech). I have discussed this rather mysterious appointment of man to succeed God in a paper called "Creation and Originality", to be published in a volume of my essays forthcoming from the Harvester Press.

18. Barbara Foxley, trans., *Emile* (London: Dent, Everyman's Library, 1966), pp. 147–48.

Gaia and the Forms of Life
STEPHEN R.L. CLARK

I

If any argument persuades the present masters of the world to preserve or, at any rate, not to wantonly destroy animal and plant species, it will probably be the economic one. The present diversity of living creatures constitutes a resource for our future. We prey upon the living world, and every time a species vanishes (irretrievably) we lose an actual or potential benefit. If we destroy the sperm whale, we may rely instead upon the joboba plant; just so, when we have exhausted fossil fuels we may cultivate ipilipil.[1] If we have already eliminated these as well, what shall we do? Medicines, materials, foodstuffs, experimental animals are all drawn from the single pool that is the terrestrial biosphere. In most cases we cannot produce equivalent objects of value from inorganic materials; when we can, it is often at enormous expense. There is no realistic hope of replacing the biosphere by an artificially maintained technosphere, and it would not be worth doing even if there was. It is therefore imprudent, to say the least, to throw away what we shall likely need (or cannot say that we will not need).

This argument may prove persuasive, though it is also true that individual companies or nation-states may see a greater gain in over-exploiting the resources despite the catastrophic, or at least disadvantageous, eventual outcome. In the absence of an enforceable code of self-restraint, what is regarded as "common property" is everyone's for the taking. Each whaling nation may calculate that if the whales will survive (as other nations exercise restraint) they may as well take advantage of the fact, and if they are doomed in any case it would be foolish not to gather a last harvest. For Hobbesian reasons it is, therefore, in all our interests to concede to some overmastering authority the right and power to prevent such overuse.

A world-wide ministry of ecological affairs would have reason to impose a very conservative strategy, keeping as much unharmed as possible, for we cannot now predict which species will turn out to hold the key to some desperate problem. But the damage done is already so severe that some discrimination may be forced on us: where so many are perishing it may prove necessary to concentrate our afforts on the safety of a few. We are living in a period which seems likely to repeat the population crashes of the Permian or the Jurassic: species may soon be vanishing at the rate of forty thousand odd a year.[2] Some will vanish because they are hunted to extinction; most because their habitats are altered for our own seeming advantage, or as the byproduct of some unthinking act. We build, irrigate, drain and deforest because we neglect the value (the mere economic value) of the habitat and its inhabitants that we thereby destroy. What would we think of early hunters who systematically slaughtered all wild horses to obtain some gastronomic delicacy or to remove a nuisance, without ever considering that there might be other "uses" for the horses, other ways of relating to them?[3] But, though we are bound to regret all our losses, it may be that we must now decide what we can least afford to lose, what we may live well enough without. Such economic calculations will be as uncertain as any other in the dismal science, resting upon judgments about where to get the resources (or energy or material) we will need to save the vanishing tribes. Sometimes, if we do not take care, we may find ourselves shooting Indian wild dogs to preserve deer and simultaneously shooting deer to preserve trees.[4] A detailed and coherent plan for the ecology of "Spaceship Earth" is something that we do not have, but may be driven into acting on. Conservatism, as we struggle to cope with the counter-intuitive results of our management techniques, will look increasingly attractive, but seem increasingly irresponsible.

It is difficult to believe that we could successfully manage the world, assigning each kind its proper place. It is especially difficult to believe, because that role will only be forced on us when so much of the world's ecosystems will have been irretrievably destroyed that we shall have no chance of finding out how a successfully managed planet actually works. It is also difficult to believe that we would be right to manage the world in the way implied by strict proponents of the economic argument. Surely our troubles stem from the assumption that the living and nonliving world is rightly regarded as a resource? It may be right to remind slave owners that they should treat their slaves as *valuable* property, if as property at all, but if that is all they are they will be treated only with such care as may seem useful to their masters at the time. The "economist" is accepting, even if only rhetorically, that we are in the right to exploit things for our own advantage, and merely reminding us that the natural world is neither infinite nor indestructible. Once the whales or the jojoba plant are gone, they are gone for good, and there is no evidence that the rate of speciation has so drastically increased as to fill the vacated niches. But are we really satisfied to say that the Amazonian jungle should be preserved merely to provide armadillos for our leprosy-research establishments (it turns out that armadillos are the only known nonhumans to be susceptible to leprosy)? Might we not feel that, once the whales are gone, at least no more of them will have to drag a whaling vessel after them with harpoons lodged in their guts?[5] "Because of their close relatedness to humans, primates are especially valuable for medical research. . . ."[6] Is Myers entirely happy that this should be a reason for preserving primates? At present, of course, it is one of the factors in our cousins' rapid decline. Would a liberationist have been happy to urge slave owners to breed their own slaves instead of relying on the capture of "wild negroes"? Certainly this would avoid the brutalities of the African slave trade, the danger of destroying the finite human resources of that continent, but it would hardly seem to touch the real argument: should human beings be treated merely as means, as resources, as material for the purposes of other human beings?

That moral respect may be owed nonhumans as well as humans is a proposition that I shall not argue here: the *onus probandi* is on those who would deny it, in the absence of any plausible reason for making such a radical disjunction.[7] It is enough for the moment to point to one economic benefit of a "natural" or "wild" habitat: that people do pay to visit or to contemplate it. Its value to the "economist" can be measured in terms of the revenue they bring, or the money that would have to be set aside to earn an equivalent return. But this cannot be its only or its principal value to the (so to call them) "tourists". They pay to see the sights for many reasons, no doubt, including reasons from which few morals can be drawn (to impress their friends or to follow the fashion), but one crucial reason must be that people, or many people, appreciate the spectacle of living nature. They do so all the more when they need not feel threatened by it, and country people can sometimes feel with good reason that townees are merely sentimentalists. It is always easier to insist that someone else has to live with the nature that we wish to contemplate at a safe distance. But though the criticism may be a fair one, it may also be right to respond that the townee is, precisely, conscious of a need that the country person, too, would feel if living nature were too far disrupted. Busy parents may feel that visiting grandparents are too sentimental about the noisy, bad-tempered, demanding young they see so rarely; it does not follow that parents do not love their children. We care for children not merely (if at all) as resources, insurance policies against old age and unemployment, nor as convenient objects of sentimental attachment: we are concerned for their good, desire their well-being. So also with living nature: we appreciate "the beauties of nature" and desire our natural environment to be in good repair.

The evolutionary causes of our concern for children and for living nature are indeed very similar. Creatures of our general kind that do *not* care for their children leave few progeny, for our young are relatively few in number and need especial care. Similarly, creatures that do not mind about the health of their environment leave few descendants, for their health and survival depends upon the health of the whole of which they are a part. To foul one's nest is foolish; to stay in a fouled nest likewise. What is foul to us may be fair to others, certainly; what is fair to us, may be ruinous to our cousins (that, after all, is our problem). But we are equipped to recognize and admire the health of an environment, and of the living creatures who inhabit and modify it. If "nature" suffers we shall suffer too: and nature has not left it to our prudential self-concern to motivate us here — we are moved directly by the sight of what, in fact, will prove (in general) of benefit or of danger to our stock.

The widespread desire to see the "beauties of nature" (and, the more we look and understand, the more it comes to seem beautiful to us) is not, then, a trivial luxury but an ingrained impulse on which our survival has in the past depended (I make no claims about the extent to which this is a matter of "instinct"). That there is widespread dislike of some environment, widespread liking for another, does not prove that the one is hurtful, the other healthy, but it does point a warning that things had better be considered carefully.

We recognize and appreciate good health in our children, our friends, our environment: we like to see things going smoothly, fitting together in a way that offers some prospect of their continuance. If the seeming health is too dependent on us, or on massive support which cannot be expected to continue or which, continuing, would have harmful side-effects, we do not think the health is genuine (though we may be glad to see the disease held at bay). If someone can only cope with life when drugged we are suspicious, though we may only count what that person needs as a *drug* if it is something relatively unusual or hard to obtain. What is a drug for one creature may be merely a necessity of life for another; one creature's poison is another's trace element (or rather may be a necessity, natural to that same creature, in the right proportions). What seems to be at the root of these discriminations is the notion of a form of life that is sustainable without massive damage to the individual or to the environment: if we need a whole chemistry industry to keep us alive, with all the concomitants of poisoned watersheds or experimental animals, we can't be very healthy — but of course it takes an entire ecosphere to keep any one of us alive. If the drugs we need were readily available, without much thought, from local vegetable gardens, they would not mark us as unhealthy: we should be healthy if we had a good supply, and that would be all there was to it. Similarly an environment can be genuinely healthy if its needs are met, without undue side effects, from the natural environment. If its "health" is maintained by careful control, artificial fertilizers, pesticides and so forth (that themselves constitute a burden on the environment), which is dependent on our continuing presence and concern (that cannot be counted on), "health" is a misnomer. We are healthy if we can reasonably expect to survive as long as most members of our kind, without massive, expensive, potentially harmful interventions. The same applies to the environment.

For this reason it is not unreasonable to feel that a world wholly domesticated, farmed, exploited would not meet our needs even if it were a practicable proposition. We want to see a healthy nature, but a genuinely healthy nature is one that can get on without our constant care. If we wish to "preserve species" it is not as domesticated or denatured zoo inmates — a point acknowledged by zoo apologists when they argue (not always convincingly) that their eventual aim is to return the animals (or their descendants) to "the wild". Even if the technosphere were to contain a regulation patch of "wilderness", a few tolerated beasts, it would not meet our requirement for a world that will continue without our assistance, that does not (in the last resort) need our hands on the tiller. It is the sheer indifference to us and to our projects displayed by the rest of the natural world that attracts us, for this indifference is our evidence that things are, after all, going smoothly, that Gaia is not yet a terminal case.

II

It is not merely improvident to radically reduce the diversity and complexity of the terrestrial biosphere (hereafter, for reasons that will become evident, called "Gaia"), it is also an offense against what is likely to be a deeply rooted admiration for a healthy, self-sustaining environment. The more sorts of creatures there are in an ecosystem, the more complex the interrelations and the more "stable" the system: not that it is, therefore, more impervious to outside interference than a simpler system (it may in fact be quite easy to wreck such a stable system), but that it will be self-sustaining under normal conditions. A mature "climax" community is characterized by a diversity of creatures who together bring about conditions that are best suited to the growth of their offspring. Succession communities bring about conditions unhelpful to their own offspring (so that they are succeeded by other kinds).[8] The climax community is, in Aristotelian (which is not to say "animistic") terms, the final cause of the earlier stages: it is here that the diverse kinds all find their niche, and because there are such climax communities there are kinds available to repopulate the area if the community should be overthrown by outside influence. A climax community cannot be planted or stocked: it takes generations for the right conditions to be brought about.[9] Not all climax communities are perfectly stable: in some cases the community rocks back and forth between one state and another, as redwoods create conditions unsuitable for their own saplings' growth and yet re-emerge as the dominant kind fairly regularly.[10]

We wish there to be, we wish to be part of, such a self-sustaining community. If it is too far disrupted it will not merely take generations to put the damage straight, but may vanish forever. Some new climax may eventually emerge, but we cannot now predict what it will be, or whether it will be hospitable to us or to our descendants. Self-sustaining communities should be left to sustain themselves. Interference, even with good intentions, may prove destructive. "All conservation of wilderness is self-defeating, for to cherish we must see and fondle, and when enough have seen and fondled, there is no wilderness left to cherish."[11] The paradox is not merely that too many eager helpers and sightseers will disrupt the ecosystem as surely as the most brutal of commercial and industrial interests. The very value of "wilderness" is that it is wild, that it owes us nothing, that it will continue without aid. It is not merely the economic arguments for conservation that seem likely to destroy the very attitudes on which the desire to conserve must rest:[12] the whole impulse to conserve and to protect is at odds with our admiration for precisely what does not need protection. If it is important to us that there be a self-sustaining community, how can we achieve that end except by sustaining it ourselves — and will it then be genuinely self-sustaining? Would "Spaceship Earth", where all processes really are dependent on our good will and intelligence, be what we wish there to be; namely, a world where everything does not depend on us?

Lovelock, the author of the "Gaia hypothesis" (that the biosphere has operated as does a living organism, modifying its own environment and so maintaining conditions suitable for its own survival), has suggested that a human population of over, say, ten thousand million would require such disruption of established systems as would either leave us as inhabitants of "the prison hulk of Spaceship Earth" or demand the death of millions to allow Gaia some chance of recovery.[13] It is perhaps some consolation that Gaia would have brought forces into play against us long before that point is reached, perhaps by insectile or microbial pests that we have carefully bred to be immune to D.D.T. or penicillin. But if Gaia really did retreat into the dust and leave us as the managers of a planet-sized chemical laboratory we should not merely be hard-pressed to survive. It is doubtful that we could. To avoid such a fate, Lovelock proposes that we be especially careful of the continental shelves: the algae *Polysiphonia fastigiata* is responsible for extracting sulphur and distributing it to the creatures of the land (via air-borne dimethyl sulphide). Another species does the same for selenium.[14]

We wish there to be, we need there to be, a world of life that can sustain itself without our constant effort, and so there is. But the world is not wholly invulnerable, any more than our own bodies can maintain their health against all rebellions of their organic parts. It would be difficult, so Lovelock guesses, to wholly unravel the sinews of Gaia. It (or she) has survived disasters in the past, even if at enormous cost: the meteor strike that finished off the dinosaurs (perhaps), the formation of Pangaea that altered ocean life and, earliest of all, the release of atmospheric oxygen by the first great polluters, plants.[15] Gaia is adept at turning "pollutants" into necessary elements. Gaia would survive, most probably, even a nuclear spasm that eliminated us. But there must come a time when so many kinds have been eliminated, so many connections broken, that Gaia no longer has the internal diversity needed to adapt to changing circumstances. We can observe the event already at a local level: if new hybrid food crops are planted in preference to the ancient kinds, a pool of necessary variation may be swiftly eliminated, and we have no recourse when the hybrids prove susceptible to disease, high winds or lack of fertilizer.

We wish there to be an independently viable nature, but there is a sense in which that desire is absurd: there can be no such nature, because we are not independent of nature. We cannot expect not to affect things, even if we do so within a system that does not depend entirely upon us. So we should wish to affect things in ways that allow for, or promote, the continued survival of the whole of which we are a part. If we did inhabit "Spaceship Earth", as a contending rabble, we should have no particular incentive to believe it necessary to maintain the kinds and habitats that happen to be here: there would be as much reason to think that the effects of a given species were deleterious as that they were beneficial. It is because we inhabit or inhere in *Gaia* that we can believe that her sinews should not be torn, that all kinds of creature play, somewhere, some part in her.

It is sometimes easy to see just how this is so. Some species are "key species", like the alligators of the Everglades whose digging operations are crucial in maintaining the particular ecosystem of which they are a part.[16] We, too, have sometimes played a crucial role in creating stable ecosystems, and can learn from those successful experiments what we ought to do (when we have to intervene). If the Sahara or the Scottish highlands are monuments to human folly, at least the English hedgerows are achievements to be glad of.[17] The English countryside is (or was) a noble work of human beings precisely because it never excluded other forms of life: rather, indeed, provided new niches for our neighbours. If people vanished from the scene it would change, moving fairly rapidly (we may plausibly suppose) to a new climax community. It is important not to think that the countryside is being kept from its "true climax", the state it would be in if we were not here: we too can be part of a self-sustaining community, and do not need to look towards "the wilderness" for examples of such beauty and balance. Accordingly, we may, at least in theory, avoid Leopold's paradox that by preserving nature we destroy what we most love: it is possible to play our proper part in a self-sustaining system, so long as we do not seek too readily to exclude others. "Shame on you who add house to house and join field to field, until not an acre remains, and you are left to dwell alone in the land."[18] Isaiah's words were, of course, directed against the rich, who set themselves against their neighbours, but it is not unreasonable to apply them here to all humans. We must not think that our kind of life is all important, even in those areas where we play a crucial role in maintaining a stable eco-system, and still less in the wilderness where we go but rarely.

To lay down that we should not interfere at all, that we should never do anything to which another species might respond, lest we upset an accidental balance, would be both unrealistic and nonideal. A nature that was as susceptible as that would not be the sort of nature we are moved to admire and care for; it would be the accidentally surviving aggregate of "Spaceship Earth". It is our recognition of system that encourages us to think it possible for there to be a healthy and self-sustaining system, and may give us confidence to be not always worrying lest we put a foot wrong. Species destruction is now on such a scale that we must do all we can to halt it, or face the future of a millenial climb back to some new climax (which will most probably not include our kind at all). But we need not accept any absolute principle to the effect that no species ever be allowed to perish from the world. What matters is the maintenance of Gaia and her constituent eco-systems, not the preservation at all costs of any single line (even our own).

III

There are those who would reject the notion that we could or should regard ourselves as parts of Gaia, on something like the same terms as any other kind. For some of us, "nature" is a thing we have outgrown, to be appreciated or transformed as we may please. It hardly matters whether this belief is expressed in merely technocratic terms by admirers of high technology, or as the conviction that we need only make the land last until the Lord returns, an event expected shortly.[19] In either case nature deserves the same scorn as merely "natural man", and value lies in transcending our material origins. The particular, largely heretical, variety of Christianity that played a role in moulding Western attitudes to the nonhuman, and uncivilized, earth, has not been the only culprit. The (self-contradictory) claim of existentialists that "man has no nature", and liberal insistence that people are all that matter (however liberating an influence that doctrine may have had in other areas), have worked together to produce the assumption that there are people, and there are things, and things count only if they count to people. Value only resides in subjects, and the only subjects commonly noticed are human ones. Value indeed lies only in the conscious choices of subjects, and cannot be imposed upon them as "objective value".

"Subjectivism" (so to call a pervasive attitude of mind) is the creed that nothing matters, nothing is of value, apart from the conscious choices of (typically) human beings. To act as a morally responsible being it is necessary only to respect that principle of choice in oneself and others. Plainly there is some subterfuge in this: for, in offering conscious choice as alone deserving of respect, subjectivists seems to run counter to their own insistence that nothing simply *is* worthy of respect. Liberal humanism becomes merely one option amongst many, and those who reject it commit no objective offence. If *anything* can decently matter to me then it is not unreasonable to prefer the destruction of the human species to the least disturbance of a cockroach.

Genuine subjectivists, who acknowledge no requirement (over and above their momentary choice) of logical coherence, or epistemological good sense, or truth, are self-exiled from the human community. Those who genuinely believe that their own consciousness, their own purposes, are of value only insofar as they are valued (by whom? what does it matter if they are valued by something that has itself no value?) are embarked upon a never-ending fall. The price of coherence is that we acknowledge something, if it is only the life of reason, as objectively valuable.

It is at this point that rationalists of all persuasions have halted: very well, it is a necessary postulate of human reason that reason be respected, but we do not similarly *need* to postulate that the world, the nonhuman, the nonrational, be respected. Their value can be taken as purely derivative, optional. It is a theme which, taken seriously, has led to a disdain not merely for the nonhuman, but also for the feminine (assumed to be deficient in genuine passionless reason) and for mere sentiment.[20] Fortunately, it can be undermined. If we really set ourselves to disown and devalue the sentiments and structure of our mere animal humanity we shall find ourselves adrift (like existentialists). Nothing is present to us as valuable except through our given natures: because we have the natures that we do, as mammals of the new recension, we esteem loyalty, respect personal property, care for our young and helpless, honour the old. Because we have the natures that we do, we are curious and imaginative and seek reasons. If we seriously say that our given natures are not to be valued, we can no longer take anything we say seriously. The life of reason is possible only to social organisms that can identify themselves and their fellows through time and can find the world worth knowing.

I do not deny that it is logically possible to hate and reject all "merely natural" values, but to do so is to end in bedlam. Wholly to reject the world is as insane an option as to vilify the divine creator of our own spirit.[21] Accordingly, it is a necessary postulate of sane endeavour that our natures not be wholly valueless: to respect reason it is necessary that we respect our humanity and so, also, our existence as living beings in a world not wholly alien to us. If the world is to be worth knowing, if our faculties are to be worth trusting, the world must be acknowledged as a value.

In sum: if we can take our reasoning selves seriously without acknowledging that the world is to be taken seriously and lovingly, it can only be because we are beings quite alien to the world we briefly inhabit (as Gnostics have imagined) — but in that case we have no grounds for thinking that a world so alien is remotely comprehensible to us. If we are not aliens, we can only take ourselves seriously by taking the web, the system, the whole of which we are a part, with similar seriousness. To despise, to reconstruct, to poison, that whole is to destroy ourselves, not only for the strictly practical reasons that I have already sketched but also because it is a self-contradiction. "When they harm the earth, they harm themselves."[22]

IV

Subjectivism can of course be given a relatively humane form: we can extend the number of conscious individuals who we are willing to take seriously as sources of value. Zoophiles and animal liberationists have taken advantage of this, or been impelled by this, to urge that we reckon nonhuman individuals worthy of our respect. When we are dealing with creatures as individuals it is entirely proper that we should respect their desires, their interests. It is not unreasonable to insist that if human individuals have "rights" then there is no acceptable reason to deny them to nonhuman individuals.[23] From that perspective it is difficult to construct any convincing argument for preserving species and habitats just as such: it is individual organisms alone that can have "rights" for only they have conscious interests. Calculation of the greater good may even require that organisms perish so that others survive in happiness; insistence on the right of individuals not to be sacrificed for the sake of others does not require that they be allowed, or helped, to have offspring. The concern of liberal individualists for the welfare of our nonhuman kin is only accidentally combined with a passion for conserving species and habitats. The holistic system I am developing in this paper may eventually be able to accommodate the demands both of habitats and of individuals.[24]

So what is the holism or objectivism that I am opposing to individualism and subjectivism? We are equipped, by evolution and/or the will of God, to perceive and to admire the health of an environment, the beauty of complex organisms. We are also equipped to perceive ourselves as set against the environment and other inhabitants of that environment. But we cannot seriously suppose that we are genuinely independent creatures, that we (or even our remote ancestors) are or ever were solid monads, having purely external relations with the rest of the world. We are not the sorts of things that could have just the properties they do, whatever properties everything else had: there are, indeed, no such things at all. What I call "myself" is a tiny fragment of a physical, chemical and biochemical system that knows no absolute boundaries. My very personhood is as an element in a socio-historical tradition, a community of conversing creatures. "I" do not come first, and then enter into contractual or epistemological relations with an external world: "I" have no being at all outside the relationships that meet in "me". To respect *myself* is to respect the nature and tradition which is refracted as "my" particular character and angle on the world.

If this be accepted — and no one except a believer in immortal, independent soul substances can well deny it[25] — it is open to us to realize that we have no better reason for thinking that "we" began at birth or conception than at the bursting of the monobloc or the insemination of the seas. No conscious memory of these events is available to "me", but for that matter no memory of this person's birth or early childhood is. With some effort it is possible to devise a language adequate to this metaphysic,[26] and the mental exercise of referring all "our" experiences and properties to It at least has merit as a meditation technique, but I shall not attempt to write the rest of this paper in King-Farlow's Ittish. Enough that individuals, which have come to seem the basic blocks of the physical and political universe, can quite readily be dissolved into waves upon the sea, honoured not as singular bearers of value, but as embodiments of some form of life which is itself a facet of the single event or substance that we call the world. To speak in myth: the world has made subsystems for itself that operate under the illusion of being separate entities so that it may form a picture of its own being. To recover from the illusion is to "remember" unity.

What can be said for and about the habitats and extinguished species of this age from the point of view of It, or of Its most magnificent subsystem (here present), Gaia herself? Much already has been: Gaia subsists in the changes and relationships of species and ecosystems. Her stability is not that of unchanging emptiness: different kinds play their parts and so depart, and we have no guarantee that the human species has any different sort of lease. "We" may perish too; even Gaia may die. What It desires or values is not information housed in these bodies, though a beautiful complexity or interrelationship seems not too implausible a guess. Instead of asking what It values it seems better to enquire what is valued in the light of It, of "our" awareness of the secret unity.

One very easy, and very dangerous, mistake would be to suppose that because "the world is one" we must therefore set about *making* it one, or that our voice is especially "nature's voice". The ministry of ecological affairs might very well be convinced that all other individuals should bow before it, because all should bow before It. It is very easy to think that the "laws of nature" are the laws we prefer, particularly if we are rich and competent. It is easy to think that if there are no genuinely independent individuals there is no harm in imposing a single pattern of behaviour on all apparent individuals. It is already, unfortunately, usual for the costs of conservation to be borne, in practice, by those least able to bear them: human hunting tribes whose livelihood it is to hunt their prey are prevented from doing so, while commercial or scientific exploitation of a "resource" is largely unchecked; human agriculturalists are displaced in favour of tigers, while the richer world subsidizes the destruction of whole habitats. Conservationists should be very attentive to the charge that their movement can easily be a tool of class and race oppression. Without a deliberate policy of self-reminder it is easy to think that someone else is to blame, easy to forget that the rich (including all of us) are the ones who consume paper (trees), space and metal, who require large-scale and inefficient agricultural practices (monocultures, cattle raising) and who are largely shielded from the consequences of their choices. The vision of earth's unity should not be a pretext for continuing our despotic rule.

In the words of the *Last Whole Earth Catalog,* "We can't put it together: it is together."[27] If the metaphysical perspective that has been suggested above has any merit to it, we do not need to create unity nor impose uniformity. All individuals, just as they are, exist as patterns of relationship, and are to be valued as bearers of a continuing tradition, respected as kindred and companions within the community. Leopold wrote of the Darwininian revolution: "This new knowledge should have given us, by this time, a sense of kinship with fellow-creatures; a wish to live and let live; a sense of wonder over the magnitude and duration of the biotic enterprise."[28]

The awakening of this sense of kinship, this making real to oneself that we exist, and can only exist, as elements within modes of a continuing community (not an aggregate or socially contracting multitude of separable individuals) is the realistic analogue of the Kantian "kingdom of ends". Its ethical imperative might be represented as follows: "So act as if your maxims had to serve at the same time as universal law for all entities that make up the world."[29] Its immediate corollary: "Take no more than your share; no more than you must to sustain the particular value that you carry for the whole." Such an injunction also carries the implication that no one way of life should be imposed on all, and steps should be taken to assist diversity if too great a uniformity exists (i.e., steps to curb the insolent power of modern technocracy, or to build up a counter-force, may be necessary expedients even for the quietest wing of ecologically conscious politics).

A further element of this holistic approach lies in the rejection (not necessarily total) of injunctions to consider only the *present* state of things. It may sometimes be useful to urge that we live "in the present", seeing all things new, not worrying over what may come, not bound by stale nightmares from our past. But the present problem of our human masses lies rather in their deracination, their ignorance of all that they have been, than in any tendency to dwell upon past errors. Conversely, to awaken to the history of what surrounds and moulds us is to understand the world a little better: to have "a sense of history" (not only written history) is an essential part of the holistic view, simply because it is only over time (remembered time) that the patterns and relationships are visible, and only over time that things have any substantial being (or semisubstantial being) at all.[30] So Leopold was right again to base his plea for conservation on such a sense of times past and yet to come: "[The crane] is the symbol of our untameable past, of that incredible sweep of millenia which underlies and conditions the daily affairs of birds and men."[31] He was, however, perhaps mistaken in one use he made of this; in supposing that "our superiority to the beasts" was shown in our capacity to love what *was*.[32] Maybe this is our privilege, but to transform an attitude of respect and awe towards the biotic enterprise (and the whole natural world in which that enterprise is embedded) into a subtle piece of self-congratulation is a dangerous game. We are not greater than the world in which we live, and have no longer or more favourable lease upon the land than others do.

So habitats and species and individuals are to be honoured as symbols of our past and hope for our future. They are not static things, to be held tightly in the shape they were. All life and being is change, but what *has* been need not always be entirely lost. If in the end we turn out to have presided over a radical reduction in diversity and complexity within Gaia we shall have sold our past for pottage. Our successors then will have to begin again the slow climb to a climax. If we are lucky (or the Lord is with us) we may yet have time to slow down our destruction of our kin, and rediscover that we share a commonwealth. If we are not, Gaia's ancient defences against usurping elements may yet prove strong.

Notes

1. Norman Myers, *The Sinking Ark* (London: Pergamon, 1979), pp. 73.
2. *Ibid.*, p. 5; An unpublished paper, J. L. Simon, "On Species Loss, Deforestation and the Absence of Data", criticizing Myers and the *Global 2000 Report to the President* (Washington: G.P.O., 1980), has persuaded me that this guess may be unduly pessimistic, but it is a possible outcome. See also J. L. Simon, *The Ultimate Resource* (Oxford: Robertson and Co., 1981).
3. James E. Lovelock, *Gaia: A New Look at the Earth* (Oxford: Oxford University Press, 1974), p. 150.
4. D. W. Ehrenfeld, *Conserving Life on Earth* (New York: Oxford University Press, 1972), p. 165.
5. C. Hollands, "Animal Welfare Year in Retrospect", in *Animals' Rights: A Symposium*, ed. David Paterson and Richard Ryder (London: Centaur Press, 1979), p. 203.
6. Myers, *Sinking Ark*, p. x.
7. See Stephen R. L. Clark, *The Moral Status of Animals* (Oxford: Clarendon Press, 1977).
8. Ehrenfeld, *Life on Earth*, p. 27.
9. *Ibid.*, p. 31.
10. *Ibid.*, p. 148. R. G. Florence, "Decline of Old Growth Redwood Forest in Relation to Some Soil Microbiological Processes", *Ecology* 46 (1976) pp.

5–64. (Florence is less sure than Ehrenfeld that hardwoods play an alternating role with redwoods.)

11. Aldo Leopold, *A Sand County Almanac* (New York: Oxford University Press, 1949), p. 101.
12. Lawrence Tribe, "Ways Not to Think about Plastic Trees", *Yale Law Journal* 83 (1974): 1315; see also, Mark Sagoff, "On Preserving the Natural Environment", *Yale Law Journal* 84 (1974): 205.
13. Lovelock, *Gaia*, p. 132.
14. *Ibid.*, p. 119.
15. *Ibid.*, p. 109.
17. Ehrenfeld, *Life on Earth*, p. 129.
18. John Black, *The Dominion of Man: The Search for Ecological Responsibility* (Edinburgh: Edinburgh University Press, 1970), p. 6; Lovelock, *Gaia*, p. 112.
19. Isa. 5:8, cited in *ibid.*, p. 61.
20. It is reported that the present United States secretary for the environment holds this attitude.
21. See Stephen R. L. Clark, "Men, Animals and Animal Behaviour" in *The Nature of the Beast: Are Animals Moral?* ed H. B. Miller and W. Williams (Oxford: Oxford University Press, 1982).
22. See Idem, "God, Good and Evil", *Proc. Aristotelian Society* (1976–77); idem, "The Absence of a Gap between Fact and Value", *Proc. Aristotelian Society,* supp. vol. 80 (1980).
23. Doug Boyd, *Rolling Thunder* (New York: Delta, 1974), pp. 51–52, cited by A. R. Drengson, "Shifting Paradigms: From the Technocratic to the Person-Planetary", *Environmental Ethics* 2 (1980): 221.
24. "Rights", of course, are not metaphysical entities mysteriously "attached" to living organisms (as some of the more scholastic discussions suggest): to speak of "rights" is to use legal metaphor for what ought to be done or not done. See further, Stephen R. L. Clark, "The Rights of Wild Things", *Inquiry* 22 (1979).
25. A. S. Gunn, "Why Should We Care about Rare Species?", *Environmental Ethics* 2 (1980), pp. 17 perhaps follows a similar line. J. Baird Callicott "Animal Liberation: A Triangular Affair", *Environmental Ethics* 2 (1980): 311 gives a good survey of the dispute between individualistic and holistic ethics.
26. These would not be the souls of the Judaeo-Christian tradition, which depend for their continued being on the breath of God.
27. John King-Farlow, *Self-Knowledge and Social Relations* (New York: Science History Publications, 1979) – a work to be strongly recommended to all who value openness and intellectual adventure.
28. *Last Whole Earth Catalog,* (Portola Inst. Inc., 1971) back cover.
29. Leopold, *Sand County*, p. 109.
30. After Immanuel Kant, *The Moral Law* (London: Hutchinson, 1948), p. 100. I hope these remarks go some way to meeting Onora O'Neill's criticisms in her review of Clark, "Moral Status", in *Journal of Philosophy* 77 (1980).
31. See, Stephen R. L. Clark, *Aristotle's Man* (Oxford: Clarendon Press, 1975).
32. Leopold, *Sand County*, p. 96.
33. *Ibid.*, p. 112.

**Part Three
Attitudes to the Natural Environment**

Western Traditions and Environmental Ethics

ROBIN ATTFIELD

John Passmore's account of Western religious and ethical traditions as they apply to nature has been accepted as authoritative by many writers on animal welfare and on environmental ethics.[1] Thus, Peter Singer who draws on Passmore's material, largely follows his assessment of Western attitudes, stressing their defects and omitting only Passmore's qualifications,[2] and Richard and Val Routley accept Passmore's account of history, although they find his ethical position too conventional (and too close to traditional Western attitudes).[3] Singer certainly employs valuable historical material independent of that which he selects from Passmore; and Val Routley suggests that the tradition of belief that humanity is a steward responsible to God for the care of the earth may support a more radical position than Passmore's.[4] However, there seems to be general agreement, among these and a number of other recent writers, that the central (and, on Singer's account, the invariable) Christian position, till recently, has been despotic and anthropocentric: it has been held that this position views everything that exists in creation as being created for the sake of humankind, and that no moral constraints on the dealings of humankind with nonhuman nature exist. The Routleys follow Passmore in noting the existence of what Passmore regards as "minority" traditions, the "stewardship" tradition just mentioned and also the tradition of "cooperation with nature" through improving the land and enhancing the beauty of landscape in accordance with natural potentials. Passmore finds in these traditions the seeds of a more satisfactory and responsible version of belief in humankind's dominion over nature, whereas the Routleys find both them, and Passmore's own ethical position unsatisfactory, and maintain the need for an environmental ethic to supersede all such positions.

I have a great deal of sympathy with such ethical conclusions as those of Singer, the Routleys and Robert Brumbaugh, another chronicler of attitudes to animals: that people have obligations with respect to many nonhumans and that the flourishing of nonhumans is of intrinsic value. However, their assessments of Western traditions and their belief in the need for a new ethic seem to me much less well grounded.[5] Before such verdicts are arrived at, a less tendentious interpretation of history than those supplied by Passmore or Singer should be sought: only when a more balanced account is available can it be decided whether a clean break with traditional morality or a new environmental ethic are called for, or whether resources exist within Western traditions for the elaboration of an ethic suited to environmental and kindred problems.

These issues require an interdisciplinary approach which combines history, theology and moral philsoophy. The earlier sections of this essay therefore, mainly concern the history of ideas; the final section, however, embodies some related environmental philosophy. In section I I will summarize criticisms that I have made elsewhere of the historical account offered by Passmore, and present related comments on the account supplied by Singer. In section II I will assess and qualify Val Routley's assessment of Western traditions, such as belief in humankind's dominion, which she, like many other writers, misconstrues, and I will attempt to clarify the logic of theistic versions of belief in stewardship. Finally, in section III, I will discuss the question of whether, as Val and Richard Routley claim, Western traditions need to be rejected or drastically revised, and whether a new ethic is necessary or possible. I will argue that the traditions of our culture offer no little hope and encouragement to advocates of an environmental ethic — and that if these traditions did not there would be slender hope of the adoption of an environmental ethic.

I

Having criticized Passmore's account of Western traditions elsewhere,[6] I shall begin by outlining the criticism which there emerges, and shall proceed to consider its bearing, and that of further historical evidence, on the account offered by Singer.

Passmore is correct in rejecting the view that the Old Testament holds, that everything is made for humankind, and maintaining that its teaching about humankind's dominion over nature is consistent with the stewardship tradition. But in maintaining that it is also consistent with a despotic interpretation, in which people may treat nature as they please, he is mistaken – partly because of the evidence of the numerous prohibitions of cruel actions (which there is no reason to treat as concerned *only* with either property or with opposition to paganism), and partly because the command to dress and keep the garden in which the first people were placed was never abrogated, but, above all, because the Hebrew notion of dominion entailed answerability to God, who was believed to care for wild creatures and the uninhabited wilderness as well as for humankind. Nature, certainly, was not regarded as sacred but, *pace* Lynn White,[7] animism has not been the only form of religion in European history which has inhibited the exploitation of nature.

The New Testament position is no different, despite Paul's stray question implying that God does not care for oxen (I Cor. 9:10–11). When actually attending to the subject of nonhuman nature (Rom. 8:21–22) Paul held that it was in travail awaiting release from decay, and participation in the liberty of the sons of God. Again, Jesus spoke of God's care for sparrows, lilies and the grass of the field, and commended work even on the sabbath to preserve oxen and asses, and the rescue of the hundredth sheep; and the fourth gospel actually represents him as saying that the good shepherd lays down his life for the sheep. The story of his cursing a barren fig tree probably originated from a parable that he told, as may also the symbolic narrative of his transferring demons from a madman into swine,[8] though readers like Augustine certainly treated these narratives as gospel. Moreover, the Epistle to the Hebrews (Heb. 10:1–18) declares animal sacrifices superfluous, and the book of Revelation has every creature praising God and the lamb (Rev. 5:13) and looks forward to a restoration of Eden (Rev. 22:2; cf. Gen. 2:9). Thus, as far as the New Testament is concerned, it is false to hold that the Christian view was that everything has been made for humankind's use, or that an arrogant or despotic attitude to nature prevailed.

As to subsequent Christian history, attitudes to nature have been much more varied than Passmore allows, as is evident from a comparative reading of Passmore and of a source which he warmly commends as a storehouse of learning, C. J. Glacken's *Traces on the Rhodian Shore.*[9]

Thus, the belief that everything was made for humankind was held by some Christians, such as Origen, Peter Lombard, Aquinas and Calvin, but was expressly rejected by others such as Augustine, Descartes, John Ray, Linnaeus and William Paley, and implicitly rejected by many others. A gentle attitude to animals, as Passmore himself has shown, was exhibited by Basil and Chrysostom, who, like Saint Francis later, believed that humans and nonhuman animals have a common origin.[10] While writing to the Manichaeans, Augustine certainly claimed as dominical the belief that there were no moral ties between humans and animals, yet at other times he held that nonhuman creatures have their own intrinsic value, though less than humans, and that their value is not to be judged by human needs or pleasures. Even Aquinas, who held that cruelty to animals was wrong chiefly because it could damage property or lead to cruelty to humans, understood the work of humankind as including the further adornment of the created earth. Passmore gives the impression that belief in cooperation with nature was absent from the centuries between the Hermetic writings and the German romantics; but Glacken had already exhibited it in the works of Basil, Ambrose and Theodoret, in the monasteries of the Benedictine rule, in particular in the teaching of Bernard of Clairvaux, and in medieval forestry and land reclamation practices, such as the building of dikes and the digging of ditches.

The task of improving the land was often, indeed, seen as the completion of God's creation and, at any rate in the cases of Basil and the others just mentioned, the associated beliefs fit the stewardship tradition, a tradition which Passmore locates as appearing among Christians not until the time of the seventeenth-century Chief Justice, Sir Matthew Hale. But Glacken finds it not only in the Bible and in the Jewish philosopher Philo, but also in the sixth-century work *The Christian Topography* of Cosmas Indicopleustes, and implicitly in Basil, Ambrose and Theodoret; and his account of the ideas behind medieval forest clearance and forest conservation suggest that it was among the motives present there. Indeed, granted also the warnings against damaging the landscape from Christian writers as disparate as Albertus Magnus, John Evelyn and Herder, and Calvin's view of the whole earth as the sphere of humankind's stewardship, there is good reason to conclude that despotism was not the central Christian tradition, and that the interrelated "minority" or "lesser" traditions of stewardship and of cooperation with nature were at least as prominent, as well as being closer to the Bible. Thus, the seeds of enlightened attitudes to nature are much deeper rooted than Passmore allows.

In the matter of the treatment of animals, however, the historical evidence of a despotic attitude is stronger, including as it does the medieval and modern practice of bullfighting, the practice on the part of Descartes and his followers of vivisection, and the efforts of various theologians (though not of Descartes himself)[11] to deny the reality of animal suffering.[12] As the treatment of animals is the chosen area of the history supplied by Singer, I shall now consider the account he gives: it amounts, in fact, to a case for the prosecution, which must be answered if the above account of Christian traditions is to be upheld.

In *Animal Liberation* and other writings Singer adopts some elements of Passmore's account and also adds emphases of his own. Thus, Singer acknowledges the Old Testament passages that require consideration for animals, but holds that the Genesis view goes unchallenged — a view in which "man is the pinnacle of creation [and] all the other creatures have been delivered into his hands...."[13] Now the Old Testament certainly authorizes meat eating; but Singer's account here is misleading about the nature of the Genesis view of dominion (see above and n. 6), and ignores the Old Testament passages (such as Ps. 104 and Job 39 and 40) that strongly conflict with the belief that everything was created for human use or benefit.

Again Singer claims that "the New Testament is completely lacking in any injunction against cruelty to animals, or any recommendation to consider their interests",[14] and that, after Plutarch, "we have to wait nearly sixteen hundred years, however, before any Christian writer attacks cruelty to animals on any ground other than that it may encourage a tendency toward cruelty to humans."[15] The moral contribution of Christianity is said to have consisted of belief in the sacredness of all (and only) human lives, animals being left out of account. But animals were not left out of account (see John 10:11, Rom. 8:19–20, Col. 1:15–20, Rev. 5:13), and Jesus, as we have seen, commended rescuing oxen and asses even on the sabbath, and spoke of God's care for sparrows. As to the period after Plutarch, besides the numerous stories associating saints with animals there is the prayer of Basil, quoted by Passmore:

And for these also, O Lord,
the humble beasts, who bear with us the heat and burden of the day,
we beg thee to extend thy great kindness of heart,
for thou hast promised to save both man and beast,
and great is thy loving-kindness, O Master.[16]

This prayer supplies a reason for kindness to beasts entirely independent of human interests. So too does Chrysostom's remark, "Surely we ought to show them [i.e., beasts] great kindness and gentleness for many reasons but, above all, because they are of the same origin as ourselves."[17]

Singer's remark about the centuries after Plutarch seems to come from one of his sources, W. E. H. Lecky's *History of European Morals from Augustus to Charlemagne*.[18] Lecky writes of Plutarch as follows: "He places the duty of kindness to animals on the broad ground of the affections, and he urges that duty *with an emphasis and a detail* to which no *adequate* parallel can, I believe, be found in the Christian writings for at least seventeen hundred years."[19] Lecky may be correct here (though not over the seventeenth and eighteenth centuries), but Singer's parallel remark, which omits the qualifications that I have italicized, cannot be squared with the facts of history, even of the slightly shorter period which it concerns. Nor can his claim, made elsewhere, that "the early Christian writers were no more ready than Aristotle [who held that all nonhuman animals exist naturally for the sake of man] to give moral weight to the lives of nonhuman animals."[20]

It could not be claimed, of course, that Basil, Chrysostom or even the saints who befriended animals, influential as they were, became the exemplars of all Christians,[21] and indeed it has rather been the rabbinical writers who have emphasized the duty of kindness to animals.[22] Yet, in at least one important respect, nonhuman animals have fared better under Christianity than under either Judaism or paganism, in that the New Testament, as we have seen, prohibited animal sacrifices. The early church abandoned such practices, as is related by Tertullian, who remarks that God has no need of them,[23] and Eusebius relates that Constantine was the first emperor to abandon them.[24]

As to the performances in the arenas of the empire involving gladiatorial contests and contests between beasts or between humans and beasts, Christians, as Singer recognizes, were forbidden to attend, on pain of excommunication.[25] The prohibition is explicit in the third-century books entitled *De Spectaculis* by Tertullian and by Cyprian,[26] and Minucius Felix gives the reason that these games were both impious and cruel.[27] When Christianity came into official favour and, despite backsliding by Constantine and others, gladiatorial contests ceased from 404 A.D.[28] Singer writes that combats with wild animals continued into the Christian era, only declining because the supply of wild animals fell away,[29] but the prohibition on attendance by Christians remained, and combats between men and beasts, having lapsed in the West with the fall of the Western empire, were finally condemned by the Council of Trullo in the East at the end of the seventh century.[30] About

combats between wild animals, Lecky remarks that the virtual disappearance in their ancient form could have had something to do with
"the softening power of Christian teaching", though he unkindly adds
the opinion that this teaching had very little effect.

Here it must be acknowledged that blood sports such as bullfighting
have probably taken place in southern Europe throughout the Christian
centuries; cockfighting is recorded in England from the twelfth century,
bear-baiting continued well into the eighteenth century there, and
bull-baiting into the nineteenth, despite the efforts of Puritans to ban it
at the time of the Commonwealth.[31] By regarding beasts as outside the
scheme of salvation, many Catholic theologians, as Lecky points out,[32]
have also regarded concern for animals as beyond their range of duty,
and this position is certainly reflected in Augustine's anti-Manichaean
writings. But Lecky grants that the place of the saints in popular
religion must have enlarged people's sympathies towards animals;[33]
whereas Singer groundlessly writes of Christianity "extinguishing for a
long, long time the spark of a wider compassion" that the worst of the
Roman attitudes had allowed. The prayers for sick animals in the
medieval Roman liturgy are a sufficient refutation of claims such as
these;[34] however, it should be added that the rejection of animal
resurrection among most Protestants has not prevented a widespread
exercising of conscience over the wrongness of cruelty to animals (see
below).

Brumbaugh is thus correct to point out that the official Christian
position from the fifth to the thirteenth century was ambivalent in
the matter of humanity's relation to other animals:[35] indeed, the
ambivalence has continued right up to the present day. Brumbaugh
bases his view of the early Christian centuries on the dubious characterization of Christianity up to the thirteenth century as largely
Neoplatonic; in place of such a shaky ground, his judgment receives
ample support from the evidence presented above.

Of Aquinas himself, whose instrumentalist attitude to animals is
undeniable,[36] Singer's accounts are even darker than history warrants.
Thus, in one place Aquinas is represented as holding that, because
animals had been placed by God under humankind's dominion,
"humans could kill nonhuman animals as they pleased so long as they
were not the property of another";[37] whereas in another place it is held
that "for Aquinas, the only sound reason for avoiding cruelty to
animals was that it could lead to cruelty to humans."[38] But cruelty to
animals often damages property, and, as Singer often points out and

Aquinas himself realized,[39] killing animals often involves cruelty. Thus, the different reasons given against killing and against cruelty would often, on Aquinas' account, apply to both and supplement each other. Indeed, the withering scorn with which Aquinas is presented by Singer does not make for impartial assessment. (Even Saint Francis receives a milder form of the same treatment, reminding Singer "of someone who, in more modern terms, is 'high'. . . .)[40]

Yet Singer's findings about Aquinas are mostly accurate. Killing animals was held by Aquinas not to be wrong in itself[41] (any more than it is by Singer at least in cases where animals lack self-consciousness and are replaced).[42] It is interesting to note here that the contrary view had been held by the heretical thirteenth-century Albigensians.[43] There again, Aquinas excludes the possibility that charity is due to animals. "Charity", he says, "extends to those things only which are made to possess the good which is eternal life."[44] Indeed, that nonhumans do not qualify for eternal life has been the standard view of Christians; fortunately, as the remarks of Chrysostom and Francis show, not all Christians have regarded this doctrine as precluding eligibility for kindly treatment.

Singer's judgment is slightly more open to doubt, however, when he maintains that in Aquinas' view the Old Testament injunctions against cruelty to animals "are not intended to spare irrational animals pain".[45] As Singer relates, Aquinas held that as far as the capacity for reason is concerned, it is indifferent how one behaves to animals, and it is in this sense that Paul denies that God cares for oxen. But Aquinas then turns to the possibility that the feelings of humans and "other animals" constitute a reason for injunctions against ill-treating them, and asserts the naturalness of pity at the sufferings of others, including nonhuman animals. Assuredly, he proceeds to argue that such pity makes for compassion for fellow humans, and that this is why it is written in the Bible (Prov. 12:10) that "the just man regardeth the life of his beasts".[46] But this argument is introduced by the phrase *proximum autem est ut* (besides). Certainly what follows this phrase, an argument relating to human needs, would have been thought necessary by Aquinas for a conclusion about justice; yet what precedes such a phrase must carry some weight on its own account, at least in explaining the point of the Old Testament ruling.[47]

Thus, Aquinas did consider the sentiment of pity to be an appropriate reaction to animal suffering that the just man would show. Yet in Singer's terms Aquinas was undoubtedly "speciesist", discriminating between creatures simply on the basis of their species, and the anthropocentrism of his instrumental view of animals certainly expresses a despotic interpretation, or rather misinterpretation, of the Biblical teaching about the dominion of humankind. (Leaving the Old Testament aside, no instrumental view is compatible with passages like I Cor. 15:40, where animals, like other terrestrial bodies, are said to have a glory of their own.) Yet he did hold, as Passmore relates, that God cares for nonrational species, if not their individual members,[48] and his recognition that beasts have sensitive but mortal souls left open the possibility that the similarities between people and other animals are of moral significance, even though the way in which he made the beasts passive to their own natures and, unlike God's rational creatures, in point of purposive activity, overstressed the dissimilarities.[49] Indeed, a number of his modern followers have been willing, like Maritain and Journet,[50] to accept duties to animals without abandoning his metaphysical framework; and in the seventeenth century it was, among others, the Scholastics who, against Cartesians such as Malebranche and Bayle, insisted on the reality of animal suffering.[51]

The seventeenth century saw a renewed consciousness of the wrongness of cruelty to animals in the writings of the sceptic Montaigne;[52] recognition among the New England clergy of the naturalness of compassion for human and nonhuman suffering (and, indeed, a law of 1641 in Massachusetts prohibiting cruelty to domestic animals);[53] John Locke's educational concern to discourage children from cruelty to animals;[54] Ralph Cudworth's belief that nonhuman animals have immaterial souls;[55] and the denial by Descartes' followers that animals can suffer at all.[56] The Cartesians, indeed, manifested more than anyone before them the despotic attitude that people may treat nonhuman animals as they please; and, though Descartes himself did not deny sensation in animals, his own practice of vivisection prepared the way for those who did. Passmore and Singer both cite a passage of La Fontaine about the remorselessness of animal experimenters among the Jansenists at Port-Royal, who, as well as tormenting animals, ridiculed those who pitied them.[57] (At least some of the bystanders seem to have thought what they witnessed improper.) In England Francis Bacon had commended vivisection, and the Royal Society practised it, despite the belief of those associates of its members, the Cambridge Platonists, in animal souls and in the omnipresent life of "plastic nature": indeed, the scientists' disregard for animal suffering reflected that cruelty towards animals which pervaded English society at this time. Yet Sir Matthew Hale held that it was part of humankind's stewardship to protect the tamer animals,[58] and soon Alexander Pope was to maintain on the same grounds that cutting open dogs was wrong because we are answerable if we mismanage "the inferior creation" which has been "submitted to our power".[59] The Cartesians also believed that their practice was authorized by the dominion of humankind,[60] but could only have maintained this by an extremely cursory and selective interpretation of scripture.

The humanitarian movement, which successfully altered attitudes and practice in matters of slavery, punishment, working conditions and also the treatment of animals originated in the eighteenth century, though it had its seventeenth-century precursors (see above). In matters of animal welfare the movement was fostered in its early stages by Christian moralists such as Locke, Wollaston, Balguy and Hutcheson, and in general by Quakers, Methodists and Evangelicals, as well as by sceptics such as Montaigne, Shaftesbury, Voltaire, Hume and Bentham.[61] Thus, Passmore's view that theological doctrines retarded the movement, while sustained by some of the evidence, is in conflict with much of the rest: for the Christian humanitarians were motivated by a profound belief in Christian charity, and in general, by their religious convictions, and not in spite of them.[62] Further examples of Christians concerned to avoid harm to animals are Leibniz,[63] whose belief in the soul-like nature of all substances and the sensitivity of the souls of animals distanced him further from the Cartesians than Brumbaugh suggests,[64] and the Anglican poet William Cowper,[65] who held that the true appreciation of nature was sullied by the detestable cruelty of blood sports. Singer is correct in observing that Genesis has often been cited in defence of meat eating, and no doubt few who cited it endeavoured to make the deaths of the animals painless. Yet, in the days before factory farming began or meat production was known to contribute to world food problems, the moral case for constraints over the consumption of meat was much weaker than it now is: now that the case has become hard to refute, Christians have been at least as ready as others to take it to heart.

The acceptance of the Darwinian theory of evolution has reinforced that belief in the kinship of humans and other animals held in the past in the Old Testament (Eccles. 3:19), by Chrysostom and Francis. The theory, as Singer remarks, is difficult to square with anthropocentrism, particularly when the interdependence of species is remarked upon.[66] It also underpins the analogies between human and animal feelings and behaviour, as Brumbaugh observes.[67] Yet it does not remove from humankind the power to modify nature for good or ill, up to the point of annihilating the life-support systems of the planet and, thus, the responsibility (whether or not before God) to care for the biosphere for the sake of its present and future inhabitants of whatever species. Hence it is not fanaticism, religious or otherwise, *pace* Singer,[68] to hold that humankind has authority over other animals, that is, the power of agents subject to moral obligations. What is objectionable is to hold that this power may be exercised by people as they please. But what Passmore calls the "minority" traditions are fully compatible with both the facts of biology and the rejection of anthropocentrism.

I am claiming, however, that these are not minority traditions. Thus the tradition which advocates cooperation with nature has been a steady influence throughout the Christian centuries, and has often been aimed at the enhancement of natural beauty or the glory of God, rather than at usefulness to humans, despite the Routleys' assumption to the contrary.[69] Except where improvement was assessed by short-term human gain, this tradition has flowed parallel or intermingled with the Biblical tradition of stewardship, a tradition both ancient and modern, and reflected in a great deal of medieval practice too. Hale and Pope, as we have seen, appealed to this tradition in matters of the treatment of animals, the one area where some Christians (i.e., the Cartesians) have held *systematically* despotic beliefs. Aquinas' beliefs were also despotic, but some of his modern followers have been able to accept duties to animals nonetheless. Though some of Augustine's polemical tracts were despotic, he opposed anthropocentrism; and his main influence on attitudes to nature seems to have lain in the acceptance among adherents of the Benedictine rule of the desirability of improving the land.

Gentleness towards animals has been encouraged among Protestants, despite Calvin's anthropocentrism, by the humanitarian movement, and in popular Catholic devotion by veneration of the saints and by prayers such as those of the medieval Roman liturgy; and such gentleness has always had its advocates among Jews. As against all this, the long history of blood sports in the West reflects an extreme form of the despotic attitude in practice, though the worst pagan excesses of the arena, including those involving animals, were forbidden to Christians and eventually banned by Christian edicts. Western attitudes to nonhuman animals, then, have been ambivalent, and far from uniformly despotic, though not uniformly gentle either. If attitudes to the land

and to the beauty and frutifulness of nature are taken into account in addition to attitudes to animals, despotic attitudes cannot fairly be held to have been characteristic of the Christian tradition. Indeed, the seeds of more enlightened attitudes have much deeper roots (as I have urged above and elsewhere)[70] than Passmore or (as I can now add) than Singer allow.

II

According to Passmore, "Christianity has encouraged man to think of himself as nature's absolute master, for whom everything that exists was designed"; and this despotic position is ascribed in particular to Calvin.[71] Val Routley calls this position the "despotic view" and contrasts it with a position which she regards (perhaps rightly) as Passmore's, the "responsible-dominion view", in which skill, knowledge and care are needed in the control of nature. Common to both views is the "dominion assumption" that "it is permissible to manipulate the whole earth and what it contains exclusively in the human interest, that the value of a natural item is entirely a matter of its value for human interests, and that all constraints on behaviour with respect to nature derive from responsibilities to other humans".[72] (Granted Calvin's belief that the whole earth is a trust from God and that no agent should, by his rapacity, deprive others who need its fruits of the benefit of them, it would be more appropriate to ascribe to him the responsible-dominion view than the despotic view.)

As Routley points out, the dominion assumption is rejected wherever it is held that nonhumans are a proper object of moral concern, so long as the reason for this does not reduce to human interests. (Thus the dominion assumption can be rejected without adopting the beliefs, criticized by Passmore, in animal rights or in the sacredness of nature.) Mainstream Western traditions recognize the possibility of obligations *with respect to* items, which are not strictly *owed* to those items; and, to the extent that they include nonhuman animals among such items, they are already at odds with the dominion assumption. In particular, Routley is aware that in some versions of the stewardship tradition humans have obligations to care for the plants and animals of the earth because God cares for them, and not because of any benefit to humans. Thus, Passmore is held to be wrong in treating the stewardship tradition as a precursor of the responsible-dominion view (which can take into account the interests of future humans), rather than as a "religious expression of a less man-centred ethic" representing a *rejection* of the standard "human-centred Western view" (Routley's italics).[73]

Routley is largely right here in both logic and history, though Passmore is probably correct to find in the stewardship tradition a "seed" of his own view, for this tradition has sometimes, as in Calvin's case, been understood anthropocentrically. But more often it has been treated in a contrary sense; and Routley could cite, in accordance with section I of the essay, the Old and New Testaments and at least Basil, Chrysostom, Hale, Ray, Locke, Linnaeus, and most of the Protestants and sceptics of the humanitarian movement from the seventeenth century onwards, as also modern Catholics like Maritain and Journet, and Jews of diverse periods, such as Philo, Maimonides and Baumgarten.[74] Similarly, the related "minority" tradition of improving the land by cooperating with nature has not always treated the interests of present and future humans as the sole criterion of improvement, though an anthropocentric variant of it was held by Aquinas. But no such restriction is apparent in the writings of Basil, Ambrose or Theodoret, or in the medieval concern to improve and to conserve the land (as recorded by Glacken), or in the early modern believers in the providential character of human skills, Ray and Derham.[75] It should also be recalled that Augustine, an admirer of human improvements to the earth, rejected an anthropocentric theory of value and, thus, at least in theory, the dominion assumption.

The major point, indeed, over which Routley's assessment as so far introduced is dubious is her view that a human-centred view *was* the standard Western one. Even Descartes rejected the view that everything was made for humankind; and, despite the influence of those who adopted that view (Origen, Peter Lombard, Aquinas, Calvin), the instrumentalist view of the value of nonhumans implicit in the dominion assumption has not been characteristic of Christian liturgies or, at any rate in matters of conservation, of Christian practice. Belief in stewardship has never been altogether eclipsed among Christians, and it is to be doubted whether most of them have adhered to anthropocentrism, at least in the form required by the dominion assumption. Such a position has also been rejected by many sceptics (Montaigne, Voltaire, Hume, Bentham) and characteristically by Jewish thinkers: so, despite its influential Aristotelian and Stoic origins, the claim that it has been the standard Western position is difficult to substantiate or credit.

There is a related, logical point: what Routley calls the dominion assumption is an inaccurate characterization of the Judaeo-Christian belief in humankind's dominion over nature. The dominion assumption involves an instrumentalist view of the value of nonhumans in a kind which many (perhaps most) Jews and Christians have eschewed. Yet the stewardship view, which many (perhaps most) of them have upheld, is nevertheless an interpretation (and plausibly the only one in keeping with the Bible) of belief in humankind's dominion. Common to theistic believers in stewardship is the belief that people have a responsibility before God to care for and/or enhance the beauty and fruitfulness of the earth: this responsibility is seen as the discharge of the power and the skills entrusted by God to their hands. It is assumed, then, that humans have been endowed with considerable power over the earth and their fellow creatures, power which can be used for good or ill. If this were not held, the obligations of stewardship could not be taken to arise. Certainly the limitations of this power are more evident now than in recent centuries, but the fact that it is not guaranteed to remain in our hands, we use it, casts no doubt on its reality. This power and these responsibilities are held by most theists to be in keeping with God's purposes; and it is in their existence on this basis that the doctrine of humankind's dominion consists. It is mistaken, therefore, to read into it extraneous and unrelated claims such as that the sole criterion to be employed in discharging these responsibilities is that of human interests; unless, that is, the fallacious claim that valuers must judge by their own interests only, well criticized by the Routleys elsewhere.[76] is adopted. Belief in humankind's dominion need not involve the dominion assumption, and frequently (perhaps usually) has not done so. (Perhaps the belief that it somehow must involve the dominion assumption arises from importing the despotic connotations of the Roman concept of *dominium* rather than taking account of Hebrew ideas about the answerability of rulers to God.) Some Christians, it should be acknowledged, have adhered to misinterpretations of this Biblical belief, holding that there are few (if any) constraints upon the exercise of humankind's power over the environment; only in this way has the despotic view been a possible one. Among those, it should be added, who have adhered to a recognizable form of belief in humankind's dominion, some have adhered to versions of the steward-ship tradition or the tradition of cooperation with nature, compatible with the dominion assumption, and at least as many others to versions of these traditions that are incompatible with it. Hence, the claims made above about the joint predominance of these "minority", or "lesser", traditions: hence, also, the claim that the seeds of more enlightened ecological attitudes are deeper rooted than Passmore or Singer or (it can now be added) Val Routley suppose.

Routley, for her part, distinguishes in a different way between versions of the stewardship tradition which do not imply the dominion assumption and ones which can or do imply it. "It is only", she writes, "if God is taken as a superhuman (so that concern for these non-human items can be reduced to a matter of human or super-human interest) that the stewardship view will sanction the Dominion Assumption. On the most plausible and obvious interpretations it would not."[77] I doubt whether the stewardship view will sanction the dominion assumption only in these circumstances; for on Routley's interpretation of Passmore,[78] who sees the stewardship view as a "seed" of his own, care for plants and animals is ultimately justified by human interests (including the avoidance of bad traits of character such as a spirit of destruction), and can thus be reconciled with the dominion assumption. (I sympathize with Routley's criticism of this reductionist account, to the effect that the moral defect is only to be explained by reference to evils perpetrated on nonhumans, evils of moral import in themselves: but Passmore could reply without absurdity by denying that any explanation needs to be given.)

But the more important issue is whether there is any reason that adhering to the stewardship view in conjunction with theistic belief should sanction the dominion assumption. History, as we have seen, seems on balance to suggest otherwise. Now it is certainly a possible position in value theory, as pointed out by Frankena, to hold that what is valuable is God's will alone, and that love of one's neighbour, etc., is of value only because God desires it.[79] But, as he recognizes, it could instead be held by a theist that some items are of value for independent reasons, or even that in some way God is glorified when that which is of value comes about or is preserved, without the fundamental reason for its value being the will of God at all. Frankena does not sufficiently spell out the latter alternative, but it is not difficult to do so. Most moral theologians accept that action willed by God is right for reasons independent of being willed by God. Similarly, states of affairs favoured by God could be of value not because of his favour, but in themselves. If they come about (or are preserved) God would be glorified, and their coming about (or preservation) through human agency could sometimes be motivated by a desire to glorify God or to love what God loves. It is consistent with this account to hold that God loves what is of value because of its value, rather than conferring value on it through loving it. Indeed, this theory is the obvious counterpart of the standard account of the relation of God's will to moral rightness. Now if the stewardship view is held in conjunction with this way of understanding the relationship between God's will and value, there is no more inclination to hold that what count are human and divine interests or purposes alone, than there is to hold that people should care for plants and animals only because of human interests. Indeed there is probably less inclination: if God's concern is believed to be extended to grass and sparrows, and this is taken to imply that God recognizes that grass and sparrows have value in themselves, then, unless a special theory is held which makes this value dependent on the promotion of human interests, the intrinsic value of grass and sparrows will be acknowledged.

Routley is, I believe, right to maintain that the stewardship view would not sanction the dominion assumption "on the most plausible and obvious interpretations". My claim is that, characteristically, this remains the case when the plausible and obvious interpretations take the form of a theistic version of the stewardship view. Admittedly, Calvin's theistic version of stewardship was held together with the dominion assumption, but this was not so with the theistic versions of stewardship held either by the Biblical writers, by Basil, Chrysostom, Hale, Derham or Woolman; and the latter versions seem, in this respect at least, to have been consistent ones. (I have also maintained that at least one nontheistic version of the stewardship view — Passmore's — can be reconciled with the dominion assumption, but I should agree with Routley that the resulting position is of low plausibility.) The Routleys themselves make a point which supports this claim, when they remark that "a society's religion *expresses* its values", and that, by classifying values as religious, one has not thereby explained them away.[80] They add, helpfully, that religious views can be misrepresented if they are thought of as regarding the world, if not as human, then as "suprahuman" property, but then, unnecessarily, themselves represent the stewardship view as essentially regarding the world in just this way.[81] Stewards certainly are in most cases responsible to owners, but if creation consists of bodies each with their own glory (I Cor. 15:40), it cannot be regarded merely as expendable resources or as disposable property. Most adherents to the stewardship view have implicitly accepted that intrinsic value is to be found among nonhumans as well as among humans; this granted, stewards of the earth should be seen not only as managers of resources, but equally as curators of treasures or as trustees of the biosphere. The property metaphor suggests that nature is regarded solely as instrumental; but on the stewardship view it has characteristically also been regarded as of value in itself.

III

In "Human Chauvinism and Environmental Ethics" Val and Richard Routley present various criticisms of Western ethical traditions, "lesser" ones included, and urge the need for a new environmental ethic. In this final section I shall consider the criticisms of the "lesser" traditions, and add some remarks about proposals for a new start in ethics. (This is not the place, however, for a discussion of their detailed normative and meta-ethical theories.)

At the beginning of their critique of Western traditions the Routleys assert that, according to the dominant Western view, "nature is the dominion of man and he is free to deal with it as he pleases (since — at least on the mainstream Stoic-Augustinian view — it exists only for his sake)".[82] Such a view they rightly represent as inconsistent with an environmental ethic in which people are not free to do as they please with nature. But, quite apart from the erroneous ascription of this theory to Augustine, this passage makes belief in the dominion of humankind correspond to the despotic view, and makes the despotic view (rather, even, than the dominion assumption) the dominant Western tradition. But this view is adjusted in the pages following, where it is recognized that the "dominant position" may have been supplanted by a modification, which corresponds to what Val Routley designates the "responsible dominion view". As I have indicated above, it is doubtful that even this view, committed as it is to the dominion assumption, has historically been dominant: but it is certainly widespread for all that.

The "modified dominant view" is now explicated in terms of a deontic principle (D) and an axiological principle (A) characteristic of the Western superethic. Principle D seems to conflate and perhaps confuse a principle governing obligation with a principle governing actions which are to be met with toleration rather than coercion:[83] but my primary concern is rather with principle A, namely:

> Only those objects which are of use or concern to humans (or persons), or which are the product of human (or person) labour

or ingenuity, are of value; thus, these are all that need to be taken into account in determining best choice or best course of action, what is good, etc.

For, apart from the parentheses allowing "persons" to be substituted for "humans", principle A is a clear embodiment of the dominion assumption.

The Routleys proceed to present seven examples designed to disclose the inadequacy of both the modified dominant position and the "lesser" traditions;[84] and I should at once declare that as far as the modified dominant position" (and hence the dominion assumption) is concerned, I believe them to be successful. In each case actions that would increasingly be recognized as wrong are permitted or authorized by the modified dominant position. Thus, it is wrong for (i) the *last person* to eliminate, as far as is possible, every living thing, or (ii) the *last people* to eliminate all wild creatures and wildernesses for their own benefit. (To this topic of total use I shall return.) It is obviously wrong for (iii) the last person, under the guise of the *great entrepreneur,* to increase gross world product and despoliate nature by manufacturing automobiles for which there are no consumers, even if they are recycled after a short period; and it is plausibly wrong for (iv) that person to, similarly, maximize production for the benefit of the last people, as consumers comprising the *industrial society,* whose every need for transport and food is met by that person's automated factories and farms. (Moral intuitions here may waver, until it is reflected that the benefits of these depredations of nature include built-in obsolescence and the flesh of animals that had been deprived of natural fulfilments, and that none of this activity is intended to enrich the lives of coming generations.) It is also wrong to (v) eliminate *vanishing species* such as the blue whale, (vi) maximize food production by means of a *factory farm,* or (vii) destroy by logging a natural forest or *wilderness* and the animals that inhabit it. Even if these actions would not be wrong in all imaginable circumstances, they remain wrong unless very great evils indeed are otherwise unavoidable; this alone is enough to confute the modified dominant position, in which they are not wrong at all.

But for what reasons do the Routleys think that their examples disclose the inadequacy of the "lesser" traditions? Their reasons are as follows.[85] The "lesser" traditions have not been adequately characterized; but in any case they do not go far enough because they imply policies of complete interference with the earth's surface or of its total use. An environmental ethic would preserve some parts of the earth's surface from substantial human interference "whether of the 'improving' sort or not", whereas both the "lesser" traditions would, in fact, prefer to see the earth's surfaces shaped along the lines of the tame but ecologically impoverished European small farm and village landscape. The tradition of cooperation in particular, it is held, seeks to perfect nature, treating usefulness for human purposes as the test of perfection; the stewardship tradition assigns humankind the role, comparable to that of farm manager, of making nature productive, though natural resources are not deliberately degraded. Both positions are thus ones that fall within the shallow ecology movement as characterized by Arne Naess, and are "typically exploitative, even if only in the longer term".[86]

Now on Naess' characterization, shallow ecology movements are concerned only with the medium-term interests of humans in the developed world. But neither the stewardship view nor the cooperation view preclude concern for the needs and welfare of the people of the Third World or of the future and, as far as I am aware, their adherents have never been so parochial as to be preoccupied with only rich humans of the next two or three generations. Not even Calvin's version of the stewardship view falls foul of this criticism.

Later in their paper the Routleys recognize that stewardship might be interpreted as management that extends beyond human interests and mere resource conservation, and is aimed at fostering and preserving whatever is of intrinsic value.[87] If so, it is inconsistent to represent the stewardship view as *precluding* concern for nonhuman creatures. Historically, in any case, it has usually embodied such concern, and is thus immune to the main charges mentioned above. John Ray, certainly, admired the landscape of rural England,[88] but, as he rejected an anthropocentric theory of value, he can hardly be accused of wanting all the wonders of the sublunary creation which he celebrated to be reshaped in that mould. Nor need the successors of his tradition. The tradition of cooperation with nature, as has been pointed out above, has often been aimed at the enhancement of natural beauty or the glory of God: Basil, one of its early adherents, taught the moral significance of nonhuman animals, and it is difficult to envisage Bernard of Clairvaux, for all of his pride at the humanly contrived improvements to the land of his monastery, favouring the wholesale transformation of terrestrial nature, expressive as it was of the divine purposes and glory,[89] for the sake of the purposes and glory of humankind. It is true that there exist thinkers who believe that the inevitability and the desirability of progress lead to the total use of the planet by and for humankind but few, if any, are adherents to theistic religions which accept God's care for nonhuman creatures; and it is a strange notion of cooperation with nature, that it is permissible to extinguish all wild zones and species which lack instrumental value for humans.

The related traditions of stewardship and of cooperation with nature thus survive the examples which bring out the flaws in the modified dominant position and the dominion assumption. The *last person* and the *last people* wrongly destroy nonhuman creatures and their habitats, as also do the *great entrepreneur* and his customers, the *industrial society*; and the loss of flourishing nonhuman lives shows it to be wrong to eliminate *vanishing species,* run a *factory farm,* or clearfell a wilderness. The possibility that the total use of the Earth's surface (such as the *last people* practise) will be enjoined by any value theory on which nonhumans (or states of their's) have intrinsic value, but the value of humans (or of self-conscious animals, or their states) is greater, a possibility which the Routleys believe to be certain of realization under the pressure of increasing human populations, which would need to take over the habitats of nonhumans.[90] Further, Jesus taught that humans were of much greater value than sparrows, and neither the stewardship nor the cooperation tradition has been committed at any time to interspecies equality; so these traditions could still be open to the "total-use" objection.

It should be stressed that the objection is not one of discrimination along species boundaries. The same objection can be made with equal cogency to a theory such as Singer's (see section I) in which there is no discrimination between creatures capable of self-consciousness, whether human or nonhuman, but all of them are given priority, at least in matters of life and death, over conscious creatures which lack self-consciousness, even though the pleasures of the latter are granted intrinsic value. Indeed, the objection does not forfeit its plausibility even if it is held, against Singer, that intrinsic value attaches to many nonconscious items.

Against theories such as Singer's, John Rodman has objected that nonhumans are mistakenly treated as inferior or defective humans, whose experiences and activities are only recognized as valuable to the extent that they mirror human ones. This criticism, which I have discussed elsewhere,[91] is endorsed by the Routleys in their passage about the total use objection; but it is not a criticism to which all theories of differential value are vulnerable. For, if Singer's position is modified so that the value of the fulfilment of a creature's genetically endowed potentials is recognized, the otherness of nonhuman species is duly respected, and value is acknowledged in activities that have no human counterpart. But this does not require equal value to be accorded to the development of any natural potential, and is compatible with a higher value being found in states of creatures capable of self-consciousness, rationality, etc.

The total-use objection, then, does not turn on the treatment of nonhumans as inferior humans, nor do all "greater-value" theories regard them in that light. It turns on the lesser value of the nonhuman (or non-self-conscious) occupant of a niche for which there is (or soon will be) a human (or a self-conscious) competitor, and asserts that on "greater-value" theories the human (or the self-conscious) creature must always be allocated the niche, even if creatures of lesser value are eliminated as a consequence. One reply to this concerns ecological balance and the interdependence of species; long before "total use" was arrived at, the progressive elimination of forests, nitrogen-fixing bacteria, etc., would undermine planetary life-support systems, whether through decreases in the oxygen of the atmosphere, increases in carbon dioxide (with the possible greenhouse effect ensuing), the loss of nitrates in the soil, or in any of countless other ways. But the reply does not guarantee the survival of any species which could be replaced with impunity by another more useful to humans, and, if it could not be supplemented as a reply, could be taken as an acknowledgment that theories which recognize the intrinsic value of nonhumans are covertly anthropocentric, so long as they differentiate between different creatures over the degree to which they can carry intrinsic value.

There is, however, a more radical reply. As long as there is a finite ratio between the value of a creature of greater value and that of a creature of lesser value, a theory which recognizes differences in intrinsic value need not enjoin the replacement of the latter by the former, as long as a sufficient number of creatures of lesser value can occupy the niche. Now the extinction of a species guarantees that there will be no future members of that species, and thus the deaths of the last members of a species spell the loss of countless future lives, even if lives of lesser intrinsic value than that of self-conscious creatures. Indeed, the envisageable tenure on life of many nonhuman species, if not edged out through deliberate or accidental human activity or neglect, is much greater than that of humans. Thus, in most cases, the usurpation by humans of the last remaining habitats of nonhumans (with the consequent extinction of the nonhuman species in question) is likely, even on greater-value theories, to involve a nett loss of value, and to be forbidden. Indeed, these considerations constitute one important reason for restricting further increases in the human population, alongside considerations such as the misery likely to be caused both to extra people and to existing people if the human population continues to increase at current rates and population levels are not stabilized before long.

The total-use objection, then, is inconclusive with respect to greater-value theories in general and to the stewardship and cooperation with nature traditions in particular. Nor, I suggest, need those traditions involve human chauvinism, i.e., "substantially differential, discriminatory and inferior treatment [by humans of nonhumans] for which there is not *sufficient* justification",[92] though the point cannot be further elaborated here. Suffice it to say that they survive the Routleys' central criticisms and their seven counter-examples, and thus apparently constitute the possible basis of an acceptable environmental ethic.

There remains, however, their criticism that these "lesser" traditions have been insufficiently characterized, "especially when the religious backdrop is removed".[93] An easy reply might take the form of the observation that this incompleteness of characterization does not prevent the Routleys from criticizing the views in question; they are, after all, historical traditions and not theoretical constructs. Yet the Routleys may still press the question of to whom, on the stewardship view, are people responsible if the traditional theistic setting is discarded?[94]

To this question there are three possible kinds of answer. It could be replied that there is no-one *to* whom we are responsible – just as Lucretius held that life is given as a tenancy, not a freehold, but not by anyone in particular.[95] We should still be repsonsible *for* the care of many things, but our responsibilities would not be a trust. This, however, is to discard the feature of answerability, without which responsibilities may be neglected, and which is a key asset of the stewardship tradition in its theistic versions.

On a second kind of reply, we are responsible *to* just those same things *for* which we have responsibilities and in respect of which we have obligations. Thus, several Enlightenment philosophers saw us as responsible both *for* and *to* posterity.[96] This is credible where responsibilities concern present and future humans and other conscious creatures; but if, as I should accept, the flourishing of other living creatures (besides these) is of intrinsic value and can generate obligations, then the class of items to which we can be responsible cannot expand to match those with respect to which we have responsibilities; for we cannot be responsible or answerable to things lacking consciousness, any more than we can be to nature as a whole, as Passmore rightly remarks.[97] There again, it is at least cogent to suppose that whatever we are responsible *to* has rights, but Val Routley is justified in remarking that we can have obligations with respect to something without it having rights, and that it is unnecessary for an environmental ethic to ascribe rights to all those things for which we are obliged to care.[98]

The third kind of reply observes that we can intelligibly be said to have responsibilities to those with whom in any way we share responsibilities and, in this connection, the responsibility to care for the biosphere, the planet and its environs. The Enlightenment insight that (even though it has never done anything for us) we are answerable to posterity can be refurbished if it is recognized that we are responsible to the class of moral agents for the discharge of our responsibilities with respect to them and others. People's responsibility for nature is a task (and perhaps also a privilege) shared between agents who are, in many cases, strangers to each other, and between the generations: if any one group defaults, the burden on the others is unfairly made more burdensome. Future moral agents are not, of course, all conscious now, but they will be later: indeed, Joel Feinberg has argued cogently that future people have rights already.[99] Thus, it is possible and appropriate to regard ourselves as trustees, for our part of the overall task, of those agents whose own part we can retard or facilitate, whether present or future; and, some would doubtless add, the agents of the past, and particularly those who, by improvements, conservation or careful housekeeping, passed on to ourselves an inhabitable planet, sustaining creatures of great beauty, diversity and value. Whether or not that addition is in place, it is in any case appropriate here for believers to add to the class of those to whom we are responsible God the creator, whose work, in the cooperative tradition, it is for us to complete for his glory, and for whose creatures, recognized as good by their originator, the stewardship tradition enjoins us to care.

My suggestion, then, is that we are responsible as stewards and trustees to the class of moral agents, and that our responsibilities are to foster and preserve intrinsic value among both humans and nonhumans. I do not pretend that this is a complete ethical theory, nor can I offer a more detailed account of the stewardship or cooperation traditions; but I do hold that this suggestion is faithful to those traditions as they have historically developed, and that the historical existence of traditions of this sort is cause for very great encouragement to the advocates of an environmental ethic.

Now the Routleys hold that the stewardship tradition may boil down to "management to serve intrinsic values", but that, if so, intrinsic values are better introduced directly.[100] But if belief in intrinsic values is enshrined in existing traditions, it does not stand to be introduced, directly or otherwise, though it can be reasserted and expounded. If, on the other hand, there were no extant tradition accessible to us in which belief in intrinsic values (and their location among nonhumans) was found, then it is hard to see how this belief could be introduced at all.

Almost certainly such beliefs are widespread in our culture for, as the Routleys observe, their readers' moral consciousness already allows them to accept the wrongness of the actions depicted in the seven examples given above[101] and it is difficult to see how else to account for this response than by an existing belief in the presence of intrinsic value in the nonhuman realm. This belief is certainly in conflict with the assumptions of the Western superethic (D and A) as depicted by the Routleys, and these assumptions have certainly been held by many anthropocentric moralists. The Routleys therefore believe that they have found some judgments which are at odds with the dominant paradigm, and in systematizing them, believe that they are offering a new system of normative ethics.

Yet what makes these judgments possible? In accepting the wrongness of the actions depicted in the seven examples, we do not see ourselves as creating new values or as legislating for humanity. We could, of course, be borrowing judgments more characteristic of other cultures, and certainly belief in the presence of intrinsic value in the nonhuman realm is more all-pervading in some non-Western cultures than in the West. Yet we could not, with conviction, borrow or endorse (or imitate others who endorse) judgments of other cultures unless there were some basis in our existing patterns of moral thought (or perhaps a basis in moral thought as such) from which, whether by analogy or in some other way, we could move so as to credit these judgments as meriting our conviction.

Exposure to newly discovered facts could also (and in my belief largely does) account for people's increasing willingness to accept these judgments. Thus, on the one hand, ecological findings about interdependence, and, on the other hand, disclosures about the treatment of animals in factory farms, have done a good deal to mould our attitudes; as also have the increasing awareness of our kinship with non-human animals through acceptance of evolutionary theory, and of common patterns of motivation in people and other animals, brought to light by ethologists. Yet once again these facts could only be seen as relevant if prior patterns of moral thought allowed them to be. No one who advocates an esoteric ethic, bearing no relation to accepted forms of moral reasoning, will secure credibility or a hearing, whatever facts he or she may care to present.

In some senses, then, a new ethic is impossible; the most that is possible is a revised normative theory accommodating and enlarging upon accepted judgments. Such a revised normative theory, however, is to be much welcomed in the area of human responsibilities for nature: after all, there are quarters in which the modified dominant position and the dominion assumption are actually held. Yet the judgments on which a revised theory is grounded must themselves arise out of existing moral traditions; and, in the case of responsibilities for nature, as was pointed out in previous sections, there have long existed traditions encapsulating some of those beliefs in the presence of intrinsic value among nonhumans, which are now being advocated in the name of a new environmental ethic. There have also existed other relevant traditions, such as Pythagoreanism, Neoplatonism and the beliefs of the Manichaeans and the Albigensians; but the Judaeo-Christian traditions with which I have been concerned have been sufficiently continuous and sufficiently widespread to make these judgments over the Routleys' examples intelligible, rather than as anomalies, which the Routleys' account of our moral traditions make them seem.

The dominion assumption, indeed, is so far removed from Jewish and Christian belief in humankind's dominion that it may well not have figured in any *historically* "dominant position" at all: nonanthropocentric versions of the stewardship and cooperation with nature views may well have been historically just as widespread. (Indeed, beliefs about nature have also been held within Christianity that are different from any of these traditions, adding to the variety of the traditions we inherit, such as Alan of Lille's belief in nature as a semidivine power, tutor to humankind's and source of all fruitfulness.[102] The very variety of our traditions and the requirement of consistency prompt the effort to refine and revise them into an acceptable ethic; but I maintain that it is only the roots within existing traditions of belief in the presence of intrinsic value among nonhuman creatures which allow a revision insisting on a theory of value which affirms this presence to be credible.

If so, it is neither necessary nor possible to regard proposals for a new ethic, insofar as they are credible ones, as breaking with or superseding our moral traditions. Proposals for an environmental ethic, in particular, are a re-expression of long-standing themes of Western culture, albeit of themes of which most (and possibly all) other cultures have counterparts. Insofar as the proposals are credible ones, their relation to accepted moral traditions could hardly, I suggest, be otherwise.

Notes

1. John Passmore, *Man's Responsibility for Nature* (London: Duckworth, 1974). A slight change of emphasis over the need for moral and metaphysical revisions is apparent in the second edition of 1980.
2. Peter Singer, *Animal Liberation: A New Ethics for Our Treatment of Animals* (New York: Avon, 1975), chap. 5; idem, *Practical Ethics* (Cambridge: Cambridge University Press, 1979), p. 77; idem, "Not for Humans Only: The Place of Nonhumans in Environmental Issues", in *Ethics and Problems of the 21st Century,* ed. K. E. Goodpaster and K. M. Sayre (Notre Dame, Ind.: University of Notre Dame Press, 1979), pp. 192–93.
3. Val Routley, "Critical Notice of *Man's Responsibility for Nature",* *Australasian Journal of Philosophy* 53 (1975): 171–85; Richard Routley, "Is There a Need for a New, an Environmental Ethic?", *Proceedings of the 15th World Congress of Philosophy* (Varna, Bulgaria: 1973), pp. 205–10; Richard and Val Routley, "Human Chauvinism and Environmental Ethics", in *Environmental Philosophy,* ed. Don Mannison, Michael McRobbie and Richard Routley (Canberra: Research School of Social Sciences, Australian National University 1980), pp. 109–34.
4. Val Routley, "Critical Notice", p. 174.
5. Over many matters in environmental ethics Singer and the Routleys would, of course, disagree. One such disagreement surfaces in section III herein. With Robert S. Brumbaugh, "Of Man, Animals and Morals: A Brief History", in *On the Fifth Day: Animal Rights and Human Ethics,* ed. Richard Knowles Morris and Michael W. Fox (Washington, D.C.: Acropolis Books, 1978), pp. 6–25, my disagreements, besides those below over Neoplatonism and over Leibniz, concern his excessive schematization of the history of thought and his belief that phenomenology may allow us to put aside all preconceptions and arrive at the truth about animals. In the same volume Richard Knowles Morris portrays Judaeo-Christian attitudes as more anthropocentric than the facts allow, in his contribution "Man and Animals: Some Contem-

porary Problems", pp. 26–31. He also seems to forget, at pp. 28–29, that the New Testament declared all animals clean.

6. Robin Attfield, "Christian Attitudes to Nature", *Journal of the History of Ideas* 44 (1983); idem, *The Ethics of Environmental Concern* (Oxford: Blackwell, 1982).

7. Lynn White, "The Historical Roots of Our Ecological Crisis", *Science* 155, no. 37 (1967): 1203–07; republished in *The Environmental Handbook: Action Guide for the UK*, ed. John Barr (London: Ballantine, Friends of the Earth, 1971), pp. 3–16.

8. Thus, Stephen R. L. Clark, *The Moral Status of Animals* (Oxford: Clarendon Press, 1977), p. 196; cf. idem, "Christian Attitudes to Nature", sec. 1.

9. C. J. Glacken, *Traces on the Rhodian Shore* (Berkeley and Los Angeles: University of California Press, 1967). Passmore, *Man's Responsibility*, acknowledges a debt of gratitude to Glacken in the preface to the first edition, p. vii.

10. John Passmore, "The Treatment of Animals", *Journal of the History of Ideas* 36 (1975): 198. Bonaventure's account of Francis' belief is given in N. Wydenbruck, trans., and Otto Karrer, ed., *St. Francis of Assisi: The Legends and the Lauds* (London: Sheed and Ward, 1977), p. 161, in a passage cited by Andrew Linzey, *Animal Rights: A Christian Assessment of Man's Treatment of Animals* (London: S.C.M. Press, 1976), p. 103, no. 22. The persistence of Basil's and Chrysostom's advocacy of compassion for animals in the Eastern Church is illustrated by the seventh-century teaching of St. Isaac the Syrian, cited by A.M. Allchin, *The World is a Wedding: Explorations in Christian Spirituality* (London: Darton, Longman and Todd, 1978), p. 85.

11. Linzey, *Animal Rights*, p. 13; John Cottingham, " 'A Brute to the Brutes?' Descartes' Treatment of Animals", *Philosophy* 53 (1978): 551–59.

12. Passmore, "Treatment of Animals", pp. 204–18.

13. Singer, *Animal Liberation*, p. 205.

14. Ibid., p. 209.

15. Ibid., p. 211.

16. Passmore, "Treatment of Animals", p. 198; Passmore there acknowledges A. W. Moss, *Valiant Crusade* (London: Cassell, 1961), p. 5.

17. Linzey, *Animal Rights*, p. 103.

18. Singer, *Animal Liberation*, p. 211. Singer cites Lecky in the previous sentence and in the corresponding note (p. 233, n. 9). He also quotes extensively from Lecky at pp. 207–8.

19. W. E. H. Lecky, *History of European Morals from Augustus to Charlemagne*, 2 vols. (London: Longmans, Green and Co., 1913), 2:166.

20. Singer, "Not for Humans Only", p. 192. But even Lecky's claim is cast into doubt by the passage of St. Isaac the Syrian cited in no. 10, and also quoted by Allchin in Hugh Montefiore, ed, *Man and Nature* (London: Collins, 1971), p. 146.

21. Lecky, *European Morals*, 2:168–72.

22. Lecky, *European Morals*, 2:167, no. 3.

23. Tertullian, "To Scapula", in *Documents in Early Christian Thought*, ed. (Cambridge: Cambridge University Press, 1975), p. 228.

24. Eusebius, "Oration in honour of Constantine on the Thirtieth Anniversary of his Reign", in *Documents*, ed. Wiles and Santer, p. 232. In the sixth century, Gregory the Great could assume that the idea of reintroducing animal sacrifices was an absurdity which would never cross his readers' minds.

25. Singer, *Animal Liberation*, p. 210.

26. Tertullian, "De Spectaculis Liber" in *Patrologia Latina*, ed. J.P. Migne, 220

vols. (Paris: 1844–63), 1:627–62; Cyprian, "Liber De Spectaculis", ibid., pp. 779–88.

27. Minucius Felix, "Octavius", *Patrologia Latina*, 3:354.
28. Lecky, *European Morals*, 2:37.
29. Singer, *Animal Liberation*, p. 210.
30. Lecky, *European Morals*, 2:37; a fuller account may be found in H. H. Milman, *The History of Christianity from the Birth of Christ to the Abolition of Paganism in the Roman Empire*, 3 vols. (London: John Murray, 1840), 3:460. Over the rise and fall of animal contests in the West, see Brumbaugh, "Man, Animals and Morals", p. 9, for his quotation of "Venationes" from O. Seyffert, *A Dictionary of Classical Antiquities*, trans. and rev. H. Nettleship and J. E. Sandys (London: S. Sonnenschein and Co., 1891).
31. Lecky, *European Models*, 2:174–75. See further, E. S. Turner, *All Heaven in a Rage* (London: Michael Joseph, 1964), pp. 28–43.
32. Lecky, *European Models*, 2:173.
33. Ibid., p. 171.
34. C. W. Hume, *The Status of Animals in the Christian Religion* (London: Universities Federation for Animal Welfare, 1957), pp. 94–98. On the medieval attitude to animals, see also T. H. White, trans. and ed., *The Book of Beasts*, (London: Jonathan Cape, 1954) and especially the appendix, p. 247.
35. Brumbaugh, "Man, Animals and Morals", p. 11.
36. Aquinas *Summa Contra Gentiles* 3.2.112(1).
37. Singer, *Practical Ethics*, p. 77.
38. Singer, "Not for Humans Only", p. 193.
39. Aquinas *Summa Theologiae* 2.1.102 a6 ad1: "Blood was forbidden [in the eating of animals] *both to avoid cruelty,* and that they might abhor the shedding of human blood."
40. Singer, *Animal Liberation*, p. 215.
41. Aquinas *Summa Contra Gentiles* 3.2.112(12).
42. Peter Singer, "Killing Humans and Killing Animals", *Inquiry* 22 (1979): 152–53; idem, *Practical Ethics*, p. 104.
43. A. J. Denomy, "An Enquiry into the Origins of Courtly Love", *Mediaeval Studies* 6 (1944): 221–28.
44. Aquinas *Summa Theologiae* 2.2.25 a3 ad3.
45. Singer, *Animal Liberation*, p. 213; idm, "Not for Humans Only", p. 193.
46. Aquinas *Summa Theologiae* 2.1.102 a6 ad8. For Passmore's treatment of the same passage, see Passmore, "The Treatment of Animals", p. 201. My interpretation, in which animal suffering counts to some degree as a reason against cruelty, is further supported by the passage cited in n. 38, where the avoidance of cruelty is one of a doublet of coordinate considerations alongside the avoidance of shedding human blood.
47. The passage cited by Passmore (see n. 46 above), from Aquinas *Summa Contra Gentiles* 3.2.113, is apparently in conflict with the interpretation given here. But granted the evidence presented in n. 46, the summary of Aquinas' position, which Passmore cites, could be an incomplete one. Aquinas did, after all, hold that animals are not excluded from God's providence (*Summa Theologiae* 1.103 a5 ad2), and that natural law applies alike to man and beast (*Summa Theologiae* 2.1.94 a2).
48. Passmore, *Man's Responsibility*, p. 117; Aquinas *Summa Contra Gentiles* 3.2.112–13. It is hardly reasonable to read this passage, as Passmore does, as implying that God will preserve these species whatever man may do to their individual members.

49. Aquinas *Summa Contra Gentiles* ibid.
50. Passmore, "Treatment of Animals", p. 206, and n. 43 there.
51. Pierre Bayle, "Rorarius", in *Historical and Critical Dictionary* (1697), trans. and ed. Richard H. Popkin (Indianapolis and New York: Bobbs Merrill, 1965), pp. 221–31.
52. Singer, *Animal Liberation,* p. 217.
53. Norman S. Fiering, "Irresistable Compassion: An Aspect of Eighteenth Century Sympathy and Humanitarianism", *Journal of the History of Ideas* 37 (1976):204; Lawrence Stone, *The Family, Sex and Marriage in England, 1500–1800* (London: Weidenfeld and Nicolson, 1977), p. 237.
54. Brumbaugh, "Man, Animals and Morals", p. 17 and p. 25, n. 4, cites Locke's *Some Thoughts Concerning Education,* 5th ed. (London, 1705). For the advocacy of kindness to animals in children's literature of the late eighteenth century, see J. H. Plumb, "The New World of Children in Eighteenth Century England", *Past and Present* 67 (1975): 64–95.
55. Ralph Cudworth, *The True Intellectual System of the Universe,* 2nd ed. (London: 1743), p. 745.
56. Passmore, "Treatment of Animals", pp. 204–18.
57. Nicholas Fontaine, *Memoires pour servir a l'histoire de Port-Royal* (Cologne: 1738), 2:52–53, cited by Passmore, "The Treatment of Animals", p. 204, and by Singer, *Animal Liberation,* p. 221. On the Royal Society, see Turner, *All Heaven in a Rage,* pp. 45–47.
58. Sir Matthew Hale, *The Primitive Origination of Mankind* (London, 1677), sect. 4, chap. 8, p. 370, cited by Passmore, *Man's Responsibility,* p. 30 and by Glacken, *Tracos,* p. 481.
59. Alexander Pope, *The Guardian* (London), 21 May 1713, cited by Singer, *Animal Liberation,* p. 221.
60. Bayle, *Dictionary,* p. 221.
61. On Locke, see n. 54. On Wollaston, Balguy and Hutcheson (Christians all), see Passmore, "The Treatment of Animals", p. 209, which mentions efforts by evangelicals pp. 209–11. On Quakers, cf. John Wollman's statement of 1772 in *Christian Faith and Practice in the Experience of the Society of Friends* (London: London Yearly Meeting of the Society of Friends, 1960), para. 478. On Methodists, see Turner, *All Heaven in a Rage,* p. 50. Passmore, "The Treatment of Animals", mentions Montaigne p. 208, Hume pp. 208ff., and Bentham p. 211. Shaftesbury is mentioned by Fiering, "Irrestible Compassion", p. 202; Voltaire by Singer, *Animal Liberation,* p. 220.
62. Cf. the couplet from a hymn of Isaac Watts (1674-1748), an Independent, "Creatures as numerous as they be thin/ Are subject to Thy care;" from "I sing the almighty power of God", *The Baptist Hymn Book* (London: Psalms and Hymns Trust, 1962), hymn 58.
63. Immanuel Kant relates of Leibniz that, after using a tiny worm for scientific observation, he "carefully replaced it with its leaf on the tree so that it should not come to harm through any act of his" *Lectures on Ethics,* trans. Louis Infield [New York: Harper and Row, 1963], pp. 239–41). Kant himself, as Passmore points out ("The Treatment of Animals", p. 202), denied direct duties to animals but, as Glacken relates (*Traces,* pp. 530-35), abandoned, in his third *Critique,* his earlier teleological anthropocentrism.
64. Brumbaugh, "Man, Animals and Morals", p. 16.
65. William Cowper, *The Task* (Ilkley and London: Scolar Press, 1973), Bk. 3, pp. 106–9.
66. Singer, *Animal Liberation,* p. 226.
67. Brumbaugh, "Man, Animals and Morals", p. 20.

68. Singer, *Animal Liberation,* p. 226.
69. Routley and Routley, "Human Chauvinism and Environmental Ethics", pp. 113–14.
70. Attfield, "Christian Attitudes to Nature", passim.
71. Passmore, *Man's Responsibility,* p. 13.
72. Val Routley, "Critical Notice", p. 173.
73. Ibid., p. 174.
74. For Maimonides, see Passmore, *Man's Responsibility,* p. 12; for Baumgarten, see Passmore, "The Treatment of Animals", pp. 201–2.
75. Derham is accused of anthropocentrism by William Coleman, "Providence, Capitalism and Environmental Degradation", *Journal of the History of Ideas* 37 (1976):35; but Glacken, *Traces,* p. 423 gives evidence that Derham held a contrary position.
76. Richard and Val Routley, "Against the Inevitability of Human Chauvinism", in *21st Century,* ed. Goodpaster and Sayre, pp. 42–47; see also idem, "Human Chauvinism and Environmental Ethics", pp. 158–62.
77. Val Routley, "Critical Notice", p. 174.
78. Ibid., p. 175.
79. W. K. Frankena, "Ethics and the Environment", in *21st Century,* ed. Goodpaster and Sayre, p. 8.
80. Richard and Val Routley, "Social Theories, Self Management and Environmental Problems", in *Environmental Philosophy,* ed. Mannison, McRobbie and Routley, p. 324.
81. Ibid.
82. Routley and Routley, "Human Chauvinism and Environmental Ethics", p. 114.
83. Ibid., p. 117.
84. Ibid., pp. 121–25.
85. Ibid., p. 113.
86. Arne Naess, "The Shallow and the Deep, Long-Range Ecology Movement", *Inquiry* 16 (1973): 95–100.
87. Routley and Routley, "Human Chauvinism and Environmental Ethics", p. 138.
88. Coleman, "Providence", p. 31.
89. Glacken, *Traces,* p. 213.
90. Routley and Routley, "Human Chauvinism and Environmental Ethics", p. 143.
91. Robin Attfield, "Methods of Ecological Ethics", unpublished paper, Cardiff, 1980).
92. Routley and Routley, "Human Chauvinism and Environmental Ethics", p. 96.
93. Ibid., p. 113.
94. Ibid., p. 138.
95. Lucretius *De Rerum Natura* 3.971.
96. Carl Becker, *The Heavenly City of the Eighteenth-Century Philosophers* (New Haven and London: Yale University Press, 1932), pp. 119–68.
97. Passmore, *Man's Responsibility,* 2nd ed. (1980), p. xii.
98. Val Routley, "Critical Notice", p. 174.
99. Joel Feinberg, "The Rights of Animals and Unborn Generations", in *Philosophy and Environmental Crisis,* ed. William T. Blackstone (Athens Ga.: University of Georgia Press, 1974), pp. 64–66.
100. Routley and Routley, "Human Chauvinism and Environmental Ethics", p. 138.
101. Ibid., p. 126.
102. Glacken, *Traces,* pp. 216–18.

Traditional American Indian and Traditional Western European Attitudes Towards Nature: An Overview

J. BAIRD CALLICOTT

This paper will sketch (in broadest outline) the picture of nature endemic to two very different intellectual traditions: the familiar, globally dominant Western European civilization, on the one hand, and the presently beleaguered tribal cultures of the American Indians, on the other. It will argue that the world view typical of American Indian peoples included and supported an environmental ethic, while that of Europeans encouraged human alienation from the natural environment and an exploitative, practical relationship with it. This paper thus represents a romantic point of view; it argues that the North American "savages" were indeed more noble than "civilized" Europeans, at least in their outlook towards nature.

I do not enter into this discussion unaware of the difficulties and limitations which present themselves at the very outset. In the first place there is no *one* thing that can be called *the* American Indian belief system. The aboriginal people of the North American continent lived in environments quite different from one another and had culturally adapted to these environments in quite different ways. For each tribe there were a cycle of myths and a set of ceremonies, and from these materials one might abstract *for each* a particular view of nature. However, recognition of the diversity and variety of American Indian cultures should not obscure a complementary unity to be found among them. Despite great internal differences there were common characteristics which culturally united American Indian peoples. Joseph Epes Brown claims that

> this common binding thread is found in beliefs and attitudes held by
> the people in the quality of their relationships to the natural
> environment. All American Indian peoples possessed what has been
> called a metaphysic of nature; all manifest a reverence for the
> myriad forms and forces of the natural world specific to their
> immediate environment; and for all, their rich complexes of rites

and ceremonies are expressed in terms which have reference to or utilize the forms of the natural world.[1]

Writing from a self-declared antiromantic perspective, Calvin Martin has more recently confirmed Brown's conjecture:

> What we are dealing with are two issues: the ideology of Indian land-use and the practical results of that ideology. Actually, there was a great diversity of ideologies, reflecting distinct cultural and ecological contexts. It is thus more than a little artificial to identify a single, monolithic ideology, as though all Native Americans were traditionally inspired by a universal ethos. Still, there were certain elements which many if not all these ideologies seemed to share, the most outstanding being a genuine respect for the welfare of other life-forms.[2]

A second obvious difficulty bedeviling any discussion of American Indian views of nature is our limited ability to accurately reconstruct the abstract culture of New World peoples prior to their contact with (and the influence from) Europeans. Documentary records of pre-contact Indian thought simply do not exist. American Indian metaphysics existed imbedded in oral traditions. Left alone an oral culture may be very tenacious and persistent. If radically stressed, it may prove to be very fragile and liable to total extinction. Hence, *contemporary* accounts by contemporary American Indians of *traditional* American Indian philosophy are vulnerable to the charge of inauthenticity since, for several generations American Indian cultures, cultures preserved in the living memory of their members, have been both ubiquitously and violently disturbed by the transplanted European civilization.

Therefore, we ought to perhaps rely where possible upon the earliest written observations of Europeans concerning American Indian belief. The accounts of the North American "savages" by Europeans of the sixteenth, seventeenth and eighteenth centuries, however, are invariably distorted by ethnocentrism, which to the cosmopolitan twentieth-century student appears so hopelessly abject as to be more entertaining than illuminating. The written observations of Europeans who first encountered American Indian cultures provide rather an instructive record of the implicit European metaphysic. Since Indians were not loyal to the Christian religion, it was assumed that they must be mind-fully servants of Satan, and that the spirits about which they talked and the powers which their shamans attempted to direct must be so many demons from Hell. Concerning the Feast of the Dead among the Huron,

Brefeuf wrote in 1636 that "nothing has ever better pictured for me the confusion among the damned".[3] His account, incidentally, is very informative and detailed concerning the physical requirements and artifacts of this ceremony, but the rigidity of his own system of belief makes it impossible for him to enter sympathetically that of the Huron.

Reconstructing the traditional Indian attitude towards nature is, therefore, to some extent a speculative matter. On the other hand, we must not abandon the inquiry as utterly hopeless. Post-contact American Indians do tell of their traditions and their conceptual heritage. Among the best of these nostalgic memoirs is Neihardt's classic, *Black Elk Speaks,* one of the most important and authentic resources available for the reconstruction of an American Indian attitude towards nature. The explorers', missionaries', and fur traders' accounts of woodland Indians' attitudes are also useful, despite their ethnocentrism, since we may correct for the distortion of their biases and prejudices. Using these two sorts of sources, first-contact European records and transcribed personal recollections of tribal beliefs by spiritually favoured Indians, plus disciplined and methodical modern ethnographic reports, we may achieve a fairly reliable reconstruction of traditional Indian attitudes towards nature.

I

On the European side of the ledger, at first glance, an analogous pluralism appears to confound any generalizations. How different the lands, the languages, the lifestyles of, say, the Swedes and the Spanish, or the Slavic and Gaelic peoples. If anything, Europeans appear to be a more ethnically diverse, motley collection of folks than Indians. Europeans, however, have enjoyed a collective intellectual *history* which the Indians, for better or worse, have not. Moreover, for many centuries a common learned language was shared by scholars of every civilized European country. This fact alone constitutes an enormously unifying force. The intellectual history of Europe has been, to be sure, dialectical and disputatious, but the pendulum of opinion has swung between well-defined limits, and certain universally accepted assumptions have prevailed.

As the European style of thought was set by the Greeks of classical antiquity, I shall begin with them. And I shall treat modern science (i.e., modern European natural philosophy) as a continuation and extrapolation of certain concepts originating with the fifth- and fourth-century Greeks. Greek ideas about nature were remarkably rich and varied. But only some of these ideas, for historical reasons which cannot be explored in this discussion, inspired and informed modern natural philosophy. They became institutionalized in the modern Western world view. It is upon them, accordingly, that I shall especially focus.

Mythopoeic Greek cosmology had curious affinities with some of the central cosmological concepts of the American Indians. Sky and Earth (Uranus and Gaia) are represented by Hesiod in the *Theogony* as male and female parents (Father and Mother) of the first generation of gods and, either directly or indirectly, of all natural beings. Some Ionian Greeks in the city of Miletus apparently became disenchanted with traditional Greek mythology and embarked upon speculations of their own. Everything, they said, is water or air. Things change because of the struggle of the hot with the cold and of the wet with the dry. The implicit question — what is the nature of things, that out of which all things come, that into which all things are resolved? — proved to be both fascinating and fruitful. After about one hundred and fifty years of uninterrupted controversy, Leucippus and Democritus, with characteristic Ionian simplicity and force, brought this line of thought to a brilliant culmination in the atomic theory of matter. The atom was conceived by them to be an indestructible and internally changeless particle, "so small as to escape sensation". There are infinitely many of these. They have substance, that is they are solid or "full", and possess shape and relative size. All other qualities of things normally disclosed by perception exist, according to Democritus, only by "convention", not by "nature". In the terms of later philosophical jargon, character-istics of things such as flavour, odour, colour and sound were regarded as *"secondary"* qualities, the subjective effects of the "primary" qualities on the sensory patient. Complementary to the concept of the atom is the concept of the "void" — free, homogeneous, isotropic space. The atoms move haphazardly about in this space. Macroscopic objects are assemblages of atoms; they are wholes exactly equal to the sum of their parts. These undergo generation and destruction, which were conceived as the association and dissociation of the atomic parts. The atomists claimed to reduce all the phenomena of nature to this simple dichotomy: the "full" and the "empty", "thing" and "no-thing", the atom and space.

Thomas Kuhn succinctly comments that "early in the seventeenth century atomism experienced an immense revival . . . atomism was firmly merged with Copernicanism as a fundamental tenet of the 'new philosophy' which directed the scientific imagination".[4] The consolidated Newtonian world view included as one of its cornerstones the atomists' concept of free space, thinly occupied by moving particles or "corpuscles", as the early moderns called them. It was one of Newton's greatest achievements to supply a quantitative model of the regular motion of the putative material particles. These famous "laws of motion" made it possible to represent phenomena not only materially, but also mechanically.

That the *order* of nature can be successfully disclosed only by means of a quantitative description, a rational account in the most literal sense of that word, is itself, of course, an idea which originated in sixth-century Greece, with Pythagoras. Pythagoras' insight had such tremendous scientific potential that it led Plato to eulogize it as Promethean, a veritable theft from the gods of the key to the secrets of the cosmos. It was cultivated and developed by the subsequent Pythagorean school and by Plato himself in the *Timaeus*, a work which enjoyed enormous popularity during the Renaissance. Modern philosophy of nature might be oversimply but, nonetheless, not incorrectly portrayed as a merger of the Pythagorean intuition that the structure of the world order is determined according to ratio, to quantitative proportions, and the Democritean ontology of void space (so very amenable to geometrical anaylsis) and material particles. The intellectual elegance and predictive power of the Newtonian natural philosophy resulted, as Kuhn suggests, in its becoming virtually institutionalized in the nascent European scientific community. Its actual and potential application to practical matters, to problems of engineering and tinkering, made it also a popular, working picture of nature, gladly and roundly embraced by all Europeans participating in enlightenment.

Paul Santmire characterized the modern European attitude towards nature as it took root in the American soil in the nineteenth century as follows:

> Nature is analogous to a machine; or in the more popular version nature *is* a machine. Nature is composed of hard, irreducible particles which have neither color nor smell nor taste . . . beauty and value in nature are in the eye of the beholder. Nature is the dead *res extens,* perceived by the mind, which observes nature from a position of objective detachment. Nature in itself is basically a self-sufficient, self-enclosed complex of merely physical forces acting on colorless, tasteless, and odorless particles of hard, dead matter. That

is the mechanical view of nature as it was popularly accepted in the circles of the educated [white Americans] in the nineteenth century.[5]

Santmire's comments bring to our attention a complementary feature of the European world view of particular interest to our overall discussion. If no qualms were felt about picturing rivers and mountains, trees and (among the legions of Cartesians) even animals as inert, material, mechanical "objects", the line was drawn at the human mind. Democritus and later Hobbes had attempted a thorough-going and self-consistent materialism, but this intrusion of matter into the very soul of humanity did not catch on — everything else, maybe, but not the human ego.

The conception of the soul as not only separate and distinct from the body, but as essentially alien to it (i.e., of an entirely different, antagonistic nature) was also first introduced into Western thought by Pythagoras. Pythagoras conceived the soul to be a fallen divinity, incarcerated in the physical world as retribution for some unspecified sin. The goal in life for the Pythagoreans was to earn the release of the soul from the physical world upon death and to reunite the soul with its proper (divine) companions. This the Pythagoreans accomplished by several methods: asceticism, ritual purification and intellectual exercise, particularly in mathematics. This led Plato, who was more than passingly influenced by Pythagoras, to (half) joke in the *Phaedo* that philosophy is the study of death, an exercise in the disentanglement of the soul from the body. The Pythagoreans and Plato indeed inverted the concepts of life and death. In the *Cratylus*, for example, Plato alleges that the word "body" (*soma*) was derived from "tomb" (*sema*). The body is thus the tomb of the soul as well as its place of imprisonment.

The Pythagorean/Platonic concept of the soul as immortal and other-worldly, essentially foreign to the hostile physical world, has profoundly influenced the European attitude towards nature. It was not only revived in a particularly extreme form by Descartes in the seventeenth century, it became popularized much earlier in Pauline Christianity. The essential self, the part of a person by means of which he or she perceives and thinks, and in which resides virtue or vice, is not of this world, and has more in common with god(s) than with nature. If the natural world is the place of trial and temptation for the soul, if the body is the prison and the tomb of the soul, then nature must be despised as the source of all misery and corruption, a place of fear and loathing: "A joyless place where murder and vengeance dwell, and swarms of other fates — wasting diseases, putrefactions and fluxes — roam in darkness over the meadow of doom."[6] So what overall attitude to nature does modern, mainstream European natural philosophy convey? Plato accounted for the existence of kinds of individuals (i.e. species) by means of his theory of ideas. Each individual or specimen "participated", according to Plato, in a certain essence or form and it derived its specific characteristics from the form in which it participated.

The impression of the natural world conveyed by the theory of ideas is that the various species are determined by the static logico-mathematical order of the formal domain, and then the individual organisms, each with its preordained essence, are loosed into the physical arena to interact clumsily, catch-as-catch-can. Nature is thus perceived, like a roomfull of furniture, as a collection, a mere aggregate of individuals of various types, relating to one another in an accidental and altogether external fashion. This picture of the world is an "atomism" of a most subtle and insidious sort. It breaks a highly integrated functional system into separate, discreet and functionally unrelated sets of particulars. Pragmatically, approaching the world through this model — which we might call "conceptual" in contradistinction to "material" atomism — it is possible to radically rearrange parts of the landscape without the least concern for upsetting its functional integrity and organic unity. Certain species may be replaced by others (e.g., grain for wild flowers in prairie biomes) or removed altogether (e.g., predator extermination) without consequence, theoretically, for the function of the whole.

Aristotle recoiled, of course, from the other-worldliness of Plato's philosophy, both from his theory of the soul and his theory of ideas. Aristotle, moreover, was a sensitive empirical biologist and did almost as much to advance biology as a science as Pythagoras did for mathematics and harmonics. Aristotle's subsequent influence on biological thought has been immense. Biology today still bears the personal stamp of his genius, especially in the system of classification (as modified and refined by Linnaeus) of organisms according to species, genus, family, order, class, phylum and kingdom. This hierarchy of universals was not real or actual, according to Aristotle (to his everlasting credit); only individual organisms fully existed. However, Aristotle's taxonomical hierarchy (in isolation from evolutionary and ecological theory) resulted in a view of living nature which was, if that is possible, more ecologically blind than Plato's. Relations among things again are, in Aristotle's biological theory, accidental and inessential. A thing's essence is determined by its logical relations within the taxonomical schema rather than, as in ecological theory, by its working relations with other things in its environment – its trophic niche, its thermal and chemical requirements, and so on. Aldo Leopold expresses this point with characteristic bluntness: "The species of a layer [in the biotic pyramid] are alike not in where they came from or in what they look like, but rather in what they eat."[7] Evolutionary and ecological theory suggest, metaphysically, that the essences of things, the specific characteristics of species, are a function of their relations with other things. To convey a very un-Aristotelian thought in an Aristotelian manner of speech, relations are "prior" to the things related, and systemic wholes are "prior" to their component parts. A taxonomical view of the biotic world untransformed by evolutionary and ecological theory has the same ecologically misrepresentative feature as Plato's theory of forms: nature is seen as an aggregate of individuals, divided into various types, which have no functional connection with one another. And the *practical* consequences are the same. The biotic mantle may be dealt with in a heavy-handed fashion and rearranged to suit one's fancy without danger of dysfunctions. If anything, Aristotle's taxonomical representation of nature has had a more insidious influence upon the Western mind than Plato's real universals, since the latter could be dismissed, as often they were, as abstracted Olympians in a charming and noble philosophical romance, while metaphysical taxonomy went unchallenged as "empirical" and "scientific".

We should not forget another Aristotelian legacy, the natural *hierarchy,* according to which the world is arranged into "lower" and "higher" forms. Furthermore, Aristotle's teleology required that the lower forms exist for the sake of the higher forms. Since human beings are placed at the top of the pyramid, everything else exists for the sake of them.

II

The late John Fire Lame Deer, a reflective Sioux Indian, comments, straight to the point, in his biographical and philosophical narrative, *Lame Deer: Seeker of Visions,* that the whites (i.e., members of the European cultural tradition) imagine earth, rocks, water, and wind to be dead, "but [they] are very much alive."[8] In the previous section I have tried to explain in what sense nature, as the *res extensa*, is conceived as "dead" in the mainstream of European natural thought. To say that rocks and rivers are dead is perhaps misleading since what is now dead once was alive. Rather in the usual European view of things such objects are considered inert. But what does Lame Deer mean when he says that they are "very much alive"?

He doesn't explain this provocative assertion as discursively as one might wish, but he provides examples, dozens of examples, of what he calls the "power" in various natural entities. According to Lame Deer, "Every man needs a stone. . . . You ask stones for aid to find things which are lost or missing. Stones can give warning of an enemy, of approaching misfortune."[9] Butterflies, coyotes, grasshoppers, eagles, owls, deer, especially elk, and bear all talk, and possess and convey power. "You have to listen to all these creatures, listen with your mind. They have secrets to tell."[10]

It would seem that for Lame Deer the aliveness of natural entities (including stones which to most Europeans are merely "material objects", and epitomize lifelessness) means that they have a share in the same consciousness that we human beings enjoy. Granted, animals and plants (if not stones and rivers) are recognized to be "alive" by conventional European conceptualization, but they lack awareness in a mode and degree comparable to human awareness. Among the Cartesians, as we earlier mentioned, even animal behaviour was regarded as altogether automatic, resembling in every way the behaviour of a machine. A somewhat more liberal and enlightened view allows that animals have a dim sort of consciousness, but get around largely by "instinct", a concept altogether lacking a clear definition and one very nearly as obscure as the notorious occult qualities (the "soporific virtues", and so on) of the Schoolmen. Of course plants are regarded as, although alive, totally lacking in sentience. In any case, we hear that only human beings possess *self*-consciousness, that is, are aware that they are aware and can thus distinguish between themselves and everything else!

Every sophomore student of philosophy has learned, or should have, that solipsism is a redoubtable philosophical position and, corollary to that, that every characterization of other minds — human as well as nonhuman — is a matter of conjecture. The Indian attitude, as represented by Lame Deer, apparently was based upon the consideration that, since human beings have a physical body *and* an associated consciousness (conceptually hypostatized or reified as "spirit"), all other bodily things, animals, plants and, yes, even stones, were also similar in this respect. Indeed, this strikes me as an eminently reasonable assumption. I can no more directly perceive another human being's consciousness than I can that of an animal or plant. I *assume* that another human being is conscious since he or she is perceptibly very like me (in other respects) and I am conscious. To anyone not hopelessly prejudiced by the metaphysical *apartheid* policy of Christianity and Western thought generally, human beings closely resemble in anatomy, physiology, and behaviour other forms of life. The variety of organic forms themselves are clearly closely related and the organic world, in turn, is continuous with the whole of nature. Virtually all things might be supposed, without the least strain upon credence, as like ourselves, "alive" (i.e., conscious, aware, or possessed of spirit).

Lame Deer offers a brief but most revealing and suggestive metaphysical explanation:

> Nothing is so small and unimportant but it has a spirit given it by Wakan Tanka. Tunkan is what you might call a stone god, but he is also a part of the Great Spirit. The gods are separate beings, but they are all united in Wakan Tanka. It is hard to understand — something like the Holy Trinity. You can't explain it except by going back to the "circles within circles" idea, the spirit splitting itself up into stones, trees, tiny insects even, making them all *wakan* by his ever-presence. And in turn all these myriad of things which makes up the universe flowing back to their source, united in one Grandfather Spirit.[11]

This Lakota pantheism presents a conception of the world which is, to be sure, dualistic, but it is important to emphasize that, unlike the Pythagorean-Platonic-Cartesian tradition, it is not an *antagonistic* dualism, in which body and spirit are conceived in contrary terms and pitted against one another in a moral struggle. Further, and most importantly for my subsequent remarks, the pervasiveness of spirit in nature, a spirit *in everything* that is a splinter of the Great Spirit, facilitates a perception of the human and natural realms as unified and akin.

Consider, complementary to this pan-psychism, the basics of Siouan cosmogony. Black Elk rhetorically asks, "Is not the sky a father and the earth a mother, and are not all living things with feet or wings or roots their children?"[12] Accordingly, Black Elk prays, "Give me the strength to walk the soft earth, a relative to all that is!"[13] He speaks of the great natural kingdom as, simply "green things", "the wings of the air", "the four-leggeds", and "the two-legged".[14] Not only does everything have a spirit; in the last analysis all things are related together as members of one universal family, born of one father, the sky, the Great Spirit, and one mother, the Earth herself.

More is popularly known about the Sioux metaphysical vision than about those of most other American Indian peoples. The concept of the Great Spirit and of the Earth Mother and the family-like relatedness of all creatures seems, however, to have been very nearly a universal American Indian idea and, likewise, the concept of a spiritual dimension or aspect to all natural things. N. Scott Momaday remarked, " 'The earth is our mother. The sky is our father.' this concept of nature, which is at the center of the Native American world view, is familiar to us all. But it may well be that we do not understand entirely what that concept is in its ethical and philosophical implications."[15] And Ruth Underhill has written that "for the old time Indian, the world did not consist of inanimate materials. . . . It was alive, and everything in it could help or harm him."[16]

Concerning the Ojibwa Indians, who speak an Algonkian language and at the time of first contact maintained only hostile relations with the Sioux, Diamond Jenness reports:

> Thus, then, the Parry Island Ojibwa interprets his own being; and exactly the same interpretation he applies to everything around him. Not only men, but animals, trees, even rocks and water are tripartite, possessing bodies, souls, and shadows. [The Parry Island) Ojibwa, Jenness earlier details, divided the spirit into two parts — soul and shadow — though, as Jenness admits, the distinction between the soul and shadow was far from clear and frequently confused by the people themselves.] They all have a life like the life in human beings, even if they have all been gifted with different powers and attributes. Consider the animals which most closely resemble human beings; they see and hear as we do, and clearly they reason about what they observe. The tree must have a life somewhat like our own, although it lacks the power of locomotion. . . . Water runs; it too must possess life, it too must have a soul and a shadow. Then observe how certain minerals cause the neighboring rocks to decompose and become loose and friable; evidently rocks too have power, and power means life, and life involves a soul and a shadow. All things then have souls and shadows. And all things die. But their souls are reincarnated again, and what were dead return to life.[17]

Irving Hallowell has noted an especially significant consequence of the pan-spiritualism among the Ojibwa: "Not only animate properties", he writes, "but even 'person' attributes may be projected upon objects which to us clearly belong to a physical inanimate category."[18] Central to the concept of a person is the possibility of entering into social relations. Nonhuman persons may be spoken with, may be honoured or insulted, may become allies or adversaries no less than human persons.

The French fur traders and missionaries of the seventeenth century in the Great Lakes region were singularly impressed by the devotion to dreams of the savages with whom they lived. In 1648 Ragueneau, speaking of the Huron, according to Kinietz, first suggested that dreams were "the language of the souls".[19] This expression lacks precision but I think it goes very much to the core of the phenomenon. Through dreams and, most dramatically, through visions one came into direct contact with the spirits of both human and nonhuman persons, as it were, naked of bodily vestments. In words somewhat reminiscent of Ragueneau's, Hallowell comments, "It is in dreams that the individual comes into direct communication with the *atiso'kanak* the powerful 'persons' of the other-than-human class."[20] Given the animistic or pan-spiritualistic world view of the Indians, acute sensitivity and pragmatic response to dreaming makes perfectly good sense.

Dreams and waking experiences are sharply discriminated, but the theatre of action disclosed in dreams and visions is continuous with and often the same as the ordinary world. In contrast to the psychologized contemporary Western view in which dreams are images of sorts (like after-images), existing only "in the mind", the American Indian, while dreaming, experiences reality, often the same reality as in waking experience, in another form of consciousness, as it were by means of another sensory modality.

As one lies asleep and experiences people and other animals, places and so on, it is natural to suppose that one's spirit becomes temporarily dissociated from the body and moves about encountering other spirits. Or, as Hallowell says, "When a human being is asleep and dreaming his *otcatcakwin* (vital part, soul), which is the core of the self, may become detached from the body (*miyo*). Viewed by another human being, a person's body may be easily located and observed in space. But his vital part may be somewhere else."[21] Dreaming indeed may be one element in the art of American Indian sorcery ("bear-walking" among the Ojibwa). If the state of consciousness in dreams may be seized and controlled, and the phenomenal content of dreams volitionally directed, then the sorcerer may go where he wishes to spy upon his enemies or perhaps affect them in some malevolent way. It follows that dreams should have a higher degree of "truth" than ordinary waking experiences, since in the dream experience the person and everyone he meets is present in spirit, in essential self. This, notice, is precisely contrary to the European assumption that dreams are "false" or illusory and altogether private or subjective. For example, in the *Second Meditation*, Descartes, casting around for an example of the highest absurdity, says that it is "as though I were to say 'I am awake now, and discern some truth; but I do not see it clearly enough; so I will set about going to sleep, so that my dreams may give me a truer and clearer picture of the fact.' "[22] Yet this, in all seriousness, is precisely what the Indian does. The following episode from Hallowell's discussion may serve as illustration. A boy claimed that during a thunderstorm he saw a thunderbird. His elders were skeptical, since to see a thunderbird in such fashion, that is with the waking eye, was almost unheard of. He was believed, however, when a man who had dreamed of the thunderbird was consulted and the boy's description was *"verified"!*[23]

The Ojibwa, the Sioux and, if we may safely generalize, most American Indians, lived in a world which was peopled not only by human persons, but by persons and personalities associated with all natural phenomena. In one's practical dealings in such a world it is necessary to one's well-being and that of one's family and tribe to maintain good social relations, not only with proximate human persons, one's immediate tribal neighbours, but also with the nonhuman persons abounding in the immediate environment. For example, Hallowell reports that among the Ojibwa "when bears were sought out in their dens in the spring they were addressed, asked to come out so that they could be killed, and an apology was offered to them".[24]

In characterizing the American Indian attitude towards nature, with an eye to its eventual comparison to ecological attitudes and conservation values and precepts, I have tried to limit the discussion to concepts so fundamental and pervasive as to be capable of generalization. In sum, I have claimed that the typical traditional American Indian attitude was to regard all features of the environment as enspirited. These entities possessed a consciousness, reason, and volition, no less intense and complete than a human being's. The Earth itself, the sky, the winds, rocks, streams, trees, insects, birds and all other animals therefore had personalities and were thus as fully persons as other human beings. In dreams and visions the spirits of things were directly encountered and could become powerful allies to the dreamer or visionary. We may therefore say that the Indian's social circle, his community, included all the nonhuman natural entities in his locale as well as other members of his clans and tribe.

Now a most significant conceptual connection obtains in all cultures between the concept of a person, on the one hand, and certain behavioural restraints, on the other. Towards persons it is necessary, whether for genuinely ethical or purely prudential reasons, to act in a careful and circumspect manner. Among the Ojibwa, for example, according to Hallowell, "a moral distinction is drawn between the kind of conduct demanded by the primary necessities of securing a livelihood, or defending oneself against aggression, and unnecessary acts of cruelty. The moral values implied document the consistency of the principle of *mutual obligations* which is inherent in all interactions with 'persons' throughout the Ojibwa world."[25]

The implicit overall metaphysic of American Indian cultures locates human beings in a larger *social,* as well as physical, environment. People belong not only to a human community but to a community of all nature as well. Existence in this larger society, just as existence in a family and tribal context, place people in an environment in which reciprocal responsibilities and mutual obligations are taken for granted and assumed without question or reflection. Moreover, a person's basic cosmological representations in moments of meditation or cosmic reflection place him or her in a world all parts of which are united through ties of kinship. All creatures, be they elemental, green, finned, winged or legged, are children of one father and one mother. One blood flows through all; one spirit has divided itself and enlivened all things with a consciousness that is essentially the same. The world around, though immense and overwhelmingly diversified and complex, is bound together through bonds of kinship, mutuality and reciprocity. It is a world in which a person might feel at home, a relative to all that is, comfortable and secure — as one feels as a child in the midst of a large family.

> But very early in life the child began to realize that wisdom was all about and everywhere and that there were many things to know. There was no such thing as emptiness in the world. Even in the sky there were no vacant places. Everywhere there was life, visible and invisible, and every object gave us great interest to life. Even without human companionship one was never alone. The world teemed with life and wisdom, there was no complete solitude for the Lakota (Luther Standing Bear).[26]

III

I shall turn now to the claim made at the beginning of this discussion, namely, that in its practical consequences the American Indian view of nature is, on the whole, more productive of a cooperative symbiosis of people with their environment than is the view of nature predominant in the Western European tradition.

Respecting the latter, Ian McHarg writes that "it requires little effort to mobilize a sweeping indictment of the physical environment which is [Western] man's creation [and] it takes little more to identify the source of the value system which is the culprit".[27] According to McHarg the culprit is "the Judeo-Christian-Humanist view which is so unknowing of nature and of man, which has bred and sustained his simple-minded anthropocentricism".[28]

Popular ecologists and environmentalists (perhaps most notably, Rachel Carson and Barry Commoner, along with McHarg and Lynn White Jr.) have, with almost loving attention, recited a litany of environmental ills, spoken of "chlorinated hydrocarbons", "phosphate detergents", "nuclear tinkering", and "the gratified bulldozer" in language once reserved for detailing the precincts of Hell and abominating its seductive Prince. Given the frequency with which we are reminded of the symptoms of strain in the global biosphere and the apocalyptic rhetoric in which they are usually cast, I may be excused if I omit this particular step from the present argument. Let us stipulate that modern technological civilization (European in its origins) has been neither restrained nor especially delicate in manipulating the natural world.

With somewhat more humour than other advocates of environmental reform, Aldo Leopold characterized the modern Western approach to nature thus: "By and large our present problem is one of attitudes and implements. We are remodeling the Alhambra with a steam shovel, and we are proud of our yardage. We shall hardly relinquish the shovel, which after all has many good points, but we are in need of gentler and more objective criteria for its successful use."[29] As far as the historical roots of the environmental crisis are concerned, I have here suggested that the much maligned attitudes arising out of the Judeo-Christian tradition have not been so potent a force in the work of remodelling as the tradition of Western natural philosophy originating among the ancient Greeks and consolidated in modern scientific thought. At least the latter has been as formative of the cultural milieu, one artifact of which is the steam shovel itself, as the former; and together, mixed and blended, so to speak, they create a mentality in which unrestrained environmental exploitation and degradation could almost be predicted in advance.

It seems obvious (especially to philosophers and historians of ideas) that attitudes and values *do* directly "determine" behaviour by setting goals (e.g., to subdue the earth, to have dominion) and, through a conceptual representation of the world, providing means (e.g., mechanics and other applied sciences). Scepticism regarding this assumption, however, has been forthcoming. Yi-Fu Tuan in "Discrepancies Between Environmental Attitude and Behavior: Examples from Europe and China" states, "We may *believe* that a world-view which puts nature in subservience to man will lead to the exploitation of nature by man; and one that regards man as simply a component in nature will entail a modest view of his rights and capabilities, and so lead to the establishment of a harmonious relationship between man and his natural environment. But is this correct?"[30] Yi-Fu Tuan thinks not. The evidence from Chinese experience that he cites is ambiguous. Concerning European experience, he marshalls examples and cases in point of large scale transformations imposed, with serious ecological consequences, upon the Mediterranean environment by the Greeks and Romans. They were, of course, nominally pagans. And he concludes this part of his discussion with the remark that "against this background of the vast transformations of nature in the pagan world, the inroads made in the early centuries of the Christian era were relatively modest".[31] I hope that my discussion in section I of this paper will have amply explained the environmental impact of Greek and Roman civilization consistently with the general thesis that world view substantially affects behaviour! Among the Chinese before Westernization, the facts which Yi-Fu Tuan presents, indicate as many congruencies as discrepancies between the traditional Taoist and Buddhist attitude towards nature and Chinese environmental behaviour.

With respect to the question, "Do cultural attitudes and values really affect the collective behavior of a culture?" a simple deterministic model will not suffice. On the one hand, it seems incredible to think that *all* our conceptualizations and our representations of the nature of nature are, as it were, mere entertainment, sort of epiphenomena of the mind, while our actions proceed in some blind way from instinctive or genetically programmed sources. After all, our picture of nature defines our theatre of action. It defines both the possibilities and limitations which circumscribe human endeavour. On the other hand, the facts of history and everyday experience do not support any simple cause and effect relationship between a given conceptual and valuational set and what people do. My own view is that it is basic to human nature to both consume and modify the natural environment. Representations of the order of nature and the proper relationship of people to that order may have either a tempering, restraining effect on manipulative and exploitative tendencies or they may have an accelerating, exacerbating effect. They also give form and direction to these inherently human drives and thus provide different cultures with their distinctive styles of doing things. Further, it appears to me that, in the case of the predominant European mentality, shaped both by the Judeo-Christian and Greco-Roman images of nature and man, the effect was to accelerate the inherent human disposition to consume and modify surroundings. A kind of "take-off" or (to mix metaphors) "quantum leap" occurred, and Western European civilization was propelled for better or worse into its industrial, technological stage with a proportional increase in ecological and environmental distress. The decisive ingredient, the *sine qua non,* may have been the particulars of the European world view.

If the predominant traditional Chinese view of nature and humanity is, as it has been characterized by Yi-Fu Tuan, "quiescent" and "adaptive", the American Indian view of the world has been characterized in essence as "ecological", for example, by Steward Udall in the *Quiet Crisis.* In "First Americans, First Ecologists" Udall nostalgically invokes the memory of Thoreau, and attributes to his ghost the opinion that "the Indians were, in truth, the pioneer ecologists of this country".[32] To assert without qualification that the American Indians were ecologists is to say the least overly bold. Ecology is a part of biology, just as organic chemistry is a part of chemistry. It is a methodical and *quantitative* study of organisms in a contextual, functional relationship to conditions of their several ranges and habitats. Udall of course disclaims that his suggestion is that Indians were scientists. One might prefer to say that American Indians intuitively acquired an essentially ecological "outlook", "perspective", or "habit of mind". That would be roughly to say that Indians viewed nature as a matrix of mutually dependent functional components integrated systemically into an organic whole. It would suggest a kind of global or holistic viewpoint; it would also imply an acute sensitivity to the complex factors influencing the life cycles of living things.

To attribute to American Indians, on the one hand, a highly abstract conceptual *schematism* and, on the other, disinterested, systematic, disciplined and meticulous observation of minutiae is to press the romantic interpretation of American Indian thought much too far. Much of the material which I have already cited *does* indicate that both woodland and plains Indians were close students of their natural surroundings. Knowledge of animals and their ways, particularly those of utilitarian value, and knowledge of plants, especially edible and medicinal ones, are well-known and much-respected dimensions of traditional Indian cultures. The American Indian pharmacopoeia alone certainly testifies to Indian botanical acumen. My impression, nonetheless, is that the typically Indian representation of nature is more animistic and symbolic than mechanical and functional. The "rules" governing hunting and fishing seem more cast in the direction of achieving the correct etiquette towards game species than *consciously* achieving maximum sustained yield of protein "resources". Medicinal plants were sought as much for their magical, symbolic and representational virtues as for their chemical effects. Of course, in the case of hunting and fishing, proper manners *are* behavioural restraints and more often than not the outcome of their being followed – of correct social forms in respect to bear, beaver and so on, being observed – was to limit exploitation and therefore, incidentally, to achieve sustained yield.

To suggest that the Indians were (intuitive, natural, pioneer or even primitive) ecologists, in other words, strikes me as being very much like saying that Indian healers, such as Black Elk, were intuitive (etc.) physicians. Indian medicine was not at an earlier stage of development than European medicine, as if moving along the same path some distance behind. It followed a different path altogether. As Black Elk explains, "It is from understanding that power comes; and the power in the [curing] ceremony was in understanding what it meant; for nothing can live well except in a manner that is suited to the way the sacred Power of the World lives and moves."[33] The power that Black Elk employed was, *in his view,* ceremonial and symbolic – what it *meant,* not what it *did* to the patient. The cure thus was affected through symbolism, not biological mechanism.

The general American Indian world view (at least the one central part of it to which I have called attention) deflected the inertia of day-to-day, year-to-year subsistence in a way that resulted, on the average, in conservation. Conservation of resources may have been, but probably was not, a *consciously* posited goal, neither a personal ideal nor a tribal policy. *Deliberate* conservation would indeed, ironically, appear to be inconsistent with the spiritual and personal attributes which the Indians regarded as belonging to nature and natural things, since these are represented by most conservationists in the predominant Pinchot tradition as only commodities, subject to scarcity, and therefore in need of prudent "development" and "management". The American Indian posture towards nature was, I suggest, neither ecological nor conservative in the modern scientific sense, as much as it was moral or ethical. Animals, plants and minerals were treated as persons, and conceived to be coequal members of a natural social order.

My cautious claim that American Indians were neither deliberate conservationists nor ecologists in the conventional sense of these terms, but manifested rather a distinctly ethical attitude towards nature and the myriad variety of natural entities is based upon the following basic points. The American Indians, on the whole, viewed the natural world as enspirited. Natural beings therefore felt, perceived, deliberated and responded voluntarily as persons. Persons are members of a social order (i.e., part of the operational concept of a person is the capacity for social interaction). Social interaction is limited by (culturally variable) behavioural restraints, rules of conduct, which we call, in sum, good manners, morals and ethics. The American Indians, therefore, in Aldo Leopold's turn of phrase, lived in accordance with a "land ethic". This view is also maintained by Scott Momaday: "Very old in the Native American world view is the conviction that the earth is vital, that there is a spiritual dimension to it, a dimension in which man rightly exists. It follows logically that there are ethical imperatives in this matter."[34]

To point to examples of wastage — buffaloes rotting on the plains or beaver all but trapped out during the fur trade — which are supposed to deliver the coup de grace to all romantic illusions of the American Indian's reverence for nature[35] would be very much like pointing to examples of murder and war in European history and concluding that Europeans were altogether without a humanistic ethic of any sort. What is lacking is a useful understanding of the function of ethics in human affairs. Ethics bear, as philosophers point out, a normative relation to behaviour; they do not describe how people actually behave, they rather set out how people ought to behave. Therefore, people are free either to act in accordance with a given ethic or not. The fact that on some occasions some do not scarcely proves that ethics are not, on the whole, influential and effective behavioural restraints. The familiar Christian ethic has exerted a decisive influence within European civilization; it has inspired noble and even heroic deeds both of individuals and whole societies. The documented influence of the Christian ethic is not the least diminished by monstrous crimes on the part of individuals. Nor do shameful episodes of national depravity, such as the Spanish Inquisition, and genocide, as in Nazi Germany, refute the assertion that a humanistic ethic has palpably affected behaviour among members of the European civilization and substantially shaped the character of that civilization itself. By parity of reasoning, examples of occasional destruction of nature on the pre-Columbian American continent and even the extirpation of species, especially during periods of enormous cultural stress, as in the fur trade era, do not, by themselves, refute the assertion that the American Indian lived not only by a tribal ethic but by a land ethic as well, the *overall* and *usual* effect of which was to establish a greater harmony between Indians and their environment than enjoyed by their Euopean successors.

IV

This conclusion would not, perhaps, require further elaboration or defence had it not recently been specifically denied by two authors, Calvin Martin and Tom Regan. In this brief polemical eqilogue, I shall therefore undertake to defend it against their criticisms. Martin writes

> Land-use was therefore *not* so much *a moral issue* for the Indian as technique animated by *spiritual-social obligations* and understandings. . . . There is *nothing here to suggest morality;* certainly there is nothing to suggest the presumptuous, condescending extension of ethics from man-to-man to man-to-land, as the Leopoldian land ethic implies. When Indians referred to other animal species as "people" — just a different sort of person from man — they were not being quaint. *Nature was a community of such "people"* — "people" for whom man had a great deal of *genuine regard* and with whom he had a contractual relationship to protect one another's interests and fulfill mutual needs. Man and Nature, in short, were joined by compact — *not by ethical ties* — a compact predicated on *mutual esteem.* This was the essence of the traditional Indian-land relationship. [36]

As we see, Martin directly denies the major conclusion reached in the previous discussion: that the American Indian world view in its general and common characteristics incorporated an environmental ethos or fostered an ethical attitude towards land and the plants and animals belonging to the land community. But his statement is very puzzling when we analyse it closely and compare it further with some of the things he says immediately before it, for example, that one shared aspect of traditional American Indian world views was "*a genuine respect* for the welfare of other life forms . . . ; [that] aboriginal man felt a *genuine kinship* and often affection for wildlife and plant-life . . . ; [and that] wildlife were revered and propitiated not only out of fear that their favors might be withheld, although there was some sense of that, but also because they were felt to be *inherently deserving of such regard*".[37]

What more do we mean by "morality" or "ethics", one wonders, a sense of respect, kinship, affection, regard and esteem? At the risk of sounding trite, I suggest that it may be merely a "semantical difference" that is involved here. I have called these American Indian attitudes towards wildlife and plant-life, mountains and rivers, sky and earth, "moral" or "ethical" attitudes, while Martin apparently wishes to hold the terms "moral" and "ethical" in reserve (though for what he does not say). He does say, "Ethics were involved only when either party broke regulations, if even then", but he does not explain this cryptic remark or what precisely he means here by "ethics".[38] My view of what counts as an ethical attitude is more Humean, I suppose, while Martin's is, perhaps, more Kantian. Following Hume, I am willing to label "ethical" or "moral" behaviour towards nature motivated by esteem, respect, regard, kinship, affection and sympathy; Kant, on the other hand, regarded all behaviour motivated by "mere inclination" (i.e., sentiment or feeling), however unselfish, as lacking genuine moral worth. For Kant, to be counted as ethical an action must be inspired solely by unsentimental duty towards some abstract precept, some categorical imperative, issued by pure reason unsullied by any empirical content. Perhaps this exalted Kantian standard for ethical behaviour lies behind Martin's disclaimer. But only upon some special and highly technical definition of "ethics", such as Kant's, can Martin's discussion be rescued from the allegation that it is plainly incoherent, indeed, that it is blatantly self-contradictory.

That issue having been, if not settled, at least clarified, let us consider another of Martin's disavowals, namely that, more specifically, the overall traditional American Indian attitude towards nature was not akin to Aldo Leopold's land ethic. However, the evidence which Martin himself has so masterfully assembled and presented earlier in his book plainly contradicts this negative claim and in this case I can see no way to rescue his discussion from self-contradiction by semantical distinctions or by any other interpretive concessions.

To take but a small sample of Martin's very full account for purposes of illustrating the similarity of his impression of Indian attitudes towards nature and the attitudes towards nature recommended by Leopold's land ethic, consider the following excerpts:

Nature, as conceived by the traditional Ojibwa, was a *congeries of societies*: every animal, fish and plant species functioned in a *society* that was parallel in all respects to mankind's. . . . [39]

As we extend these ideas further, we come to realize that the key to understanding the Indian's role within Nature lies within the notion of *mutual obligation*: man and nature both had to adhere to a prescribed behavior toward one another. . . .[40]

According to Cree ideology . . . hunting rests on a kind of *social relationship* between men and animals. Throughout the cycle of hunting rites men emphasize their *respect* by means of symbolic expressions of their subordination to animals. . . . [41]

In view of Martin's point-blank statement that there is, in American Indian attitudes towards nature, "nothing to suggest" the precepts of "the Leopoldian land ethic", a closer look at the foundations of Leopold's land ethic than I have provided above now becomes necessary.

In the first place, Leopold takes a more Humean than Kantian approach to the concepts of ethics and morality: "It is inconceivable to me", he writes, "that an ethical relation to land can exist without *love, respect,* and *admiration* for land and a high *regard* for its value."[42] No mention is made of pure reason and unconditional duties; rather, certain sentiments — love, respect, regard, admiration — are the crucial ingredients of an ethical attitude.

Secondly, it is, according to Leopold, social membership to which ethics and ethical attitudes are correlative: "All ethics so far evolved rest upon a single premise: that the individual is a member of a community of interdependent parts."[43]

Finally, the primary feature of a land ethic is the representation of nature as a congeries of societies and human-nonhuman relationships as essentially social: "The land ethic simply enlarges the boundary of the *community* to include soils, waters, plants, and animals, or collectively: the land."[44]

As we see, according to the reconstruction that Martin himself so ably and persuasively presents, though he later unaccountably denies it, the traditional American Indian attitude towards nature was, in its shared and general assumptions, so similar to the essential notions of Leopold's land ethic as to be basically identical to it.

Finally, Martin does not explain why he finds Leopold's land ethic "presumptuous" and "condescending", nor shall I speculate on his reasons for these epithets. I find Leopold's land ethic to be, if anything, the opposite of presumptuous and condescending. For the sake of comparison, there is a very different sort of "environmental ethic", the so-called animal liberation or animal rights ethic, that is based upon an arbitrary condition, "sentience", for moral considerability, which all human beings paradigmatically exhibit, with moral standing then extended to higher "lower animals" on the grounds that they too manifest the same quality.[45] This humane ethic does seem to me to be condescending and presumptuous. In Leopold's land ethic, the *summum bonum* resides in the "biotic community" and moral value or moral standing devolves upon plants, animals, people and even soils and waters by virtue of their membership in this (vastly) larger-than-human society.[46] As Leopold rather bluntly puts it, "a land ethic changes the role of *Homo sapiens* from conqueror of the land-community to plain member and citizen of it".[47] The privileged position of human beings in the natural order is, thus, in the Leopoldian land ethic, done away with in a single bold stroke. How can this be either presumptuous or condescending?

There is another unjustifiably sceptical remark, which Calvin Martin makes, as it were, as his parting shot at the neoromantic environmentalist view of Indians. He says that

> even if we absolve him of his ambiguous culpability in certain episodes of despoliation, invoking instead *his pristine sentiments toward Nature* [once more, are these not the very soul of an ethical attitude?], the Indian still remains a misfit guru. . . . The Indian's was a profoundly different cosmic vision when it came to interpreting Nature — a vision Western man would never adjust to. There can therfore be no salvation in the Indian's traditional conception of Nature for the troubled environmentalist.[48]

This statement, though brief and scarcely defended, has had wide influence. For example, one reviewer for a distinguished journal devoted to scholarship on American Indians, though thoroughly critical of Martin's ethnohistorical methods and his controversial hypothesis of an Indian-animal war, assumes without question or criticism that "his epilogue disparages effectively contemporary views about the 'ecological' Indian".[49]

After having so sharply contrasted the traditional Western European attitude towards nature with the traditional American Indian, perhaps I should whole-heartedly agree with Martin on this particular, especially as the basis of his claim goes back to cultural roots or fundamentals. Indeed, it is precisely because Western culture is grounded, according to Martin, in the Judeo-Christian tradition that "even if he [the Indian] were capable of leading us we could not follow".[50]

A full discussion of this issue would go far beyond the scope of this section. However, in view of Martin's pessimistic conclusion, I will hazard a more optimistic suggestion. It may prove to be true that in its own fashion Western science, particularly ecology and the life sciences but also physics and cosmology, is contributing to the development of a *new* Western world view remarkably similar in *some* ways (but only in some) to that more or less common to American Indian cultures. Science in the twentieth century has retreated from its traditional mechanistic and materialistic biases; indeed, twentieth-century science has been, in just this respect, "revolutionary". Popular Western culture still lags behind. Europeans and Euro-Americans remain, for the most part, nominally Christian and unregenerately materialistic and mechanistic, but the new biocentric and organic world view (embedded in a holistic cosmology and coupled with a field-theory ontology) has already begun to emerge. And there is, further, every reason to expect that it will eventually fully flower in the form of a wholly new popular culture. Present interest in environmental pollution, endangered species, popularized ecology *and* American Indian environmental attitudes and values are all harbingers of this emerging consciousness.

By the foregoing criticisms I certainly do not intend to belittle Calvin Martin's major achievement in the main body of *Keepers of the Game*, which is a monumental contribution to the recent effort to reconstruct the outlines of the traditional American Indian outlook upon the world. Tom Regan, in a discussion which is very dependent upon Martin's, takes a step towards articulating the very real philosophical problem which Martin may have felt but was unable to either coherently state or effectively resolve. According to Regan, "there is always . . . the possibility that it was fear of the keepers [spiritual] warden's of game animals in woodland Indian metaphysics], not appreciation of nature's inherent values, that directed these People's [sic] behavior".[51] Regan therefore believes that this consideration contributes "an ineradicable layer of ambiguity to the respectful behavior of Native Peoples [towards nature]".[52]

With Regan's observations before us we can now formulate what may have been the actual basis of Martin's reluctance to call American Indian attitudes towards nature "ethical" or "moral". Indian self-interest alone may have dictated deference towards nature since, if such restraint were not forthcoming, nature, everywhere spiritually enlivened as they believed, would withhold its sustenance or, worse, actively retaliate, and the Indians would woefully suffer. Now, any pattern of behaviour motivated by mere selfishness on all accounts (including Hume's, to say nothing of Kant's) is certainly not properly described as moral or ethical. Hence, for all their touted spiritualizing of nature and reverence and restraint in taking game and gathering plants, perhaps "there is nothing here to suggest morality", to quote Martin again.

Regan quite fairly points out that "the ambiguity of Amerind behaviour is the ambiguity of human behavior, the ancient puzzle over whether, as humans, we are capable of acting out of disinterested respect for what we believe has value in its own right or whether, beneath all manner of ceremony, ritual and verbal glorification of the objects of our attention, there resides, in Kant's memorable words, 'the dear self', the true, the universal sovereign of our wills".[53] As this observation clearly suggests, to describe *any* human behaviour whatsoever as "ethical" or "moral" may be naive and incautious. Personally, I think Regan and perhaps Martin (if this is in fact the basis of his reservations) are being overly cynical about human nature. I am more inclined to agree with Hume when he says, respecting human motives of behaviour, that while

> we may justly esteem our *selfishness* the most considerable, I am sensible, that, generally speaking, the representations of this quality have been carried much too far; and that the descriptions, which certain philosophers delight so much to form of mankind in this particular, are as wide of nature as any accounts of monsters, which we meet with in fables and romances. So far from thinking that men have no affection for anything beyond themselves, I am of opinion, that tho' it be rare to meet with one, who loves any single person better than himself; yet 'tis rare to meet with one, in whom all the kind affections taken together, do not over-balance all the selfish.[54]

Human beings are not, I think, incapable of acting from motives of affection, sympathy, regard, respect, fellow-feeling, reverence and so on, as well as from purely selfish motives; thus, I think human beings *per se* are not incapable of ethical or moral behaviour.

Most folk ethics (as distinct from formal, philosophical theories) take account of and play upon both our moral and selfish sentiments. Take the familiar Christian ethic as an example. We are urged to love God and to love our neighbour as ourselves. Those whose moral sentiments "over-balance" the selfish probably do genuinely behave respectfully, for the most part, towards other persons because they love God and at least sympathize with (if not love) one another. On the other hand, there is an appeal to the dear self. If you do not obey God's commandments and at least act *as if* you respected other persons, God will punish you.

In this respect it seems to me that Indian land ethics are precisely analogous to Western humanitarian ethics. Nowhere in this discussion have I claimed that traditional American Indians were morally better than Westerners in the sense that they were more altruistic and European and Euro-Americans more selfish. Rather, I have claimed that these two broad *cultural traditions* provide very different views of nature and thus very differently excite or stimulate the moral sentiments of their members. In persons belonging to both cultures there is, we may be sure, a mixture of selfishness and altruism. The ratio does not vary so much from culture to culture as from individual to individual. Some individuals in any culture may be very nearly devoid of benevolent feelings, while others are so filled with them that they willingly sacrifice themselves for the sake of their fellows. Hence Indian land ethics, like the humanitarian religious folk ethic of Western culture, is in this respect bilateral.

In traditional American Indian cultures the animals and plants were commonly portrayed as fellow members of a Great Family or Great Society. They were "persons" worthy of respect, even affection. But if there were individuals incapable of other-oriented feelings, people who were narrowly selfish (and in every human group there always are), then an appeal to fear of punishment was included, as it were, as a backup motivation to the same forms of action that others were motivated to do because of *moral* sentiments. Noble and generous American Indians would be mortified to wantonly and needlessly slaughter game animals for sport alone, since the animals were fellow members of an extended family or extended society and had themselves been so generous and cooperative. On the other hand, meaner Indians might be induced to submit to similar restraints because of fear of retribution. This retributive factor does not suggest, to me at any rate, that American Indian *world views* did not, therefore, include a land ethic (Martin's claim), or even that there is, therefore, an "ineradicable layer of ambiguity" in the average manifestly restrained behaviour of traditional Indians towards nature (Regan's claim), an ambiguity which makes it impossible for us to decide if their restrained relations with nature were genuinely moral or merely selfish. Such restraints were doubtlessly of both the moral and selfish sorts and the balance between these two behavioural poles varied from person to person and, with respect to a given person, probably from time to time. The point is that American Indian *cultures* provided their members with an environmental ethical *ideal,* however much it may have been, from time to time or from person to person, avoided, ignored, violated or, for that matter, grudgingly honoured because of fear of punishment.

Notes

1. Joseph E. Brown, "Modes of Contemplation Through Action: North American Indians", *Main Currents in Modern Thought* 30 no. 2 (1973–74):60.
2. Calvin Martin, *Keepers of the Game: Indian Animal Relations and the Fur Trade* (Berkeley and Los Angeles: University of California Press, 1978), p. 186.
3. W. Vernon Kinietz, *Indians of the Western Great Lakes: 1615–1760* (Ann Arbor: The University of Michigan Press, 1965), p. 115.
4. Thomas S. Kuhn, *The Copernican Revolution: Planetary Astronomy and the Development of Western Thought* (Cambridge: Harvard University Press, 1957), p. 237.
5. H. Paul Santmire, "Historical Dimensions of the American Crisis", in *Western Man and Environmental Ethics,* ed. Ian G. Barbour (Menlo Park: Addison-Wesley Publishing Co., 1973), pp. 70–71.
6. Empedocles *Purifications* DK 31 B 121, trans. John Mansley Robinson, in *An Introduction to Early Greek Philosophy* (New York: Houghton Mifflin, 1968), p. 152.
7. Aldo Leopold, *A Sand County Almanac* (New York: Oxford University Press, 1949), p. 215.
8. Richard Erdoes, *Lame Deer: Seeker of Visions* (New York: Simon & Shuster, 1976), pp. 108–9.
9. Ibid., p. 101.
10. Ibid., p. 124.
11. Ibid., pp. 102–3.
12. John G. Neihardt, *Black Elk Speaks* (Lincoln: The University of Nebraska Press, 1932), p. 3.
13. Ibid., p. 6.
14. Ibid., p. 7.
15. N. Scott Momaday, "A First American Views His Land", *National Geographic* (July, 1976), p. 14.

16. Ruth M. Underhill, *Red Man's Religion: Beliefs and Practices of the Indians North of Mexico* (Chicago: The University of Chicago Press, 1965), p. 40.
17. Diamond Jenness, *The Ojibwa Indians of Parry Island, Their Social and Religious Life*, Canadian Department of Mines Bulletin 78 (1935): 20–21.
18. A. Irving Hallowell, "Ojibwa Ontology, Behavior, and World View", in *Culture in History: Essays in Honor of Paul Radin*, ed. S. Diamond (New York: Columbia University Press, 1960), p. 26.
19. Kinietz, *Indians of the Western Great Lakes*, p. 126.
20. Hallowell, "Ojibwa Ontology", p. 19.
21. Ibid., p. 41.
22. Rene Descartes, *Philosophical Writings*, ed. and trans. Elizabeth Anscombe and Peter Geach (Melbourne: Thomas Nelson, 1970), p. 70.
23. Hallowell, "Ojibwa Ontology", p. 32.
24. Ibid., p. 35.
25. Ibid., p. 47 (emphasis added).
26. Brown, "Modes of Contemplation", p. 64.
27. Ian McHarg, "Values, Process, Form", in *The Ecological Conscience*, ed. Robert Disch (Englewood Cliffs, N.J.: J. Prentice-Hall, 1970), p. 25.
28. Ibid., p. 98.
29. Leopold, *Sand County*, pp. 225–26.
30. Yi-Fu Tuan, "Discrepancies Between Environmental Attitude and Behavior", in *Ecology and Religion in History* ed. David and Eileen Spring (New York: Harper and Rowe, 1974), p. 92.
31. Ibid., p. 98.
32. Steward Udall, "First Americans, First Ecologists", in *Look to the Mountain Top*, ed. Charles Jones (San Jose: Gousha Publications, 1972), p. 2.
33. Neihardt, *Black Elk Speaks*, p. 212.
34. Udall, "First Americans", p. 18.
35. The most scurrilous example of this sort of argument with which I am acquainted is "Primitive Man's Relationship to Nature" by Daniel A. Guthrie, *Bioscience* (July 1971): 721–23. In addition to rotting buffalo, Guthrie cites alleged extirpation of pleistocene megafauna by Paleo-Indians c. 10,000 years B.P. (as if that were relevant) and his cheapest shot of all, "the litter of bottles and junked cars to be found on Indian reservations today".
36. Martin, *Keepers*, p. 187 (emphasis added).
37. Ibid., p. 186 (emphasis added).
38. Ibid., p. 187.
39. Ibid., p. 71 (emphasis added).
40. Ibid., p. 77 (emphasis added).
41. Ibid., p. 116 (emphasis added); Martin is here quoting Adrian Tanner, with approval.
42. Leopold, *Sand County*, p. 227 (emphasis added).
43. Ibid., p. 203.
44. Ibid., p. 204 (emphasis added).
45. Cf. J. Baird Callicott, "Animal Liberation: A Triangular Affair", *Environmental Ethics* 2, no. 4 (1981):311–38.
46. Cf. ibid., p. 324.
47. Leopold, *Sand County*, p. 204.
48. Martin, *Keepers*, p. 188 (emphasis added).
49. Kenneth M. Morrison (review of Martin's *Keepers*), *American Indian Culture and Research Journal* 3, no. 1 (1979):78.
50. Martin, *Keepers*, p. 188.
51. Tom Regan, "Environmental Ethics and the Ambiguity of the Native American

Relationship with Nature", in *All That Dwell Therein* (Berkeley and Los Angeles: University of California Press, 1982). Martin does, of course, mention such fears (see n. 39) but does not pursue the line of thought this psychological element suggests to Regan.

52. Ibid.
53. Ibid., p. 35.
54. David Hume, *A Treatise of Human Nature* (Oxford: The Clarendon Press, 1960), pp. 486–87.

Roles and Limits of Paradigms in Environmental Thought and Action
RICHARD ROUTLEY

Sometimes those who differ over an environmental issue, be it forest destruction or nuclear development or species disappearance, differ only in a shallow way: for example, over calculations of utilities from a destructive strategy, or over means to agreed ends. But sometimes, and increasingly often, those who differ over such an issue differ in a deeper way as to fundamental assumptions. Towards the extremes of this, as the distance between those who differ increases, lie, so it is said, paradigm differences; differences as to underlying ideology, or world view, as it might previously have been put. This is one way into the theory of paradigm differences. Another perhaps better way is this:

> Reflection on the way people argue concerning live environmental issues reveals that an argument can often be cast into the following forms.
>
> 1. From a proper environmental perspective, practice (development, method . . .) P cannot be justified;
>
> 2. But even from a conventional perspective, practice P is not justified (because, for example, when analysed it has an unfavourable cost-benefit analysis).
>
> Hence, practice P ought not to be continued.

The argument is an elaboration of the scheme: $A \rightarrow B$, $\sim A \rightarrow B$, $B \rightarrow C$ C (where $B \rightarrow C$ represents the premise that what is not justified ought not to be happening). Examples of live environmental issues that very naturally lend themselves to presentation in this sort of way are arguments concerning forestry practices (e.g., Australian woodchip projects), arguments concerning nuclear development, and arguments over dam construction. It is an attempt, in short, to render deductively tight (probably a faulty aim) such arguments as:

> Nuclear development isn't justified under environmental assumptions, it isn't really justified under the opposition's (nonenvironmental) assumptions either; so it isn't justified.

Evidently the form of the argument needs filling out to explain what the different perspective are, and to show that they have the right properties (e.g., that the competing perspectives are suitably exhaustive to carry the argument). Moreover, the argument really needs elaborating to explain why the practices commonly persist *despite* such arguments (of a decisive sort). But such elaboration can be given: there are, at least in outline, sociological explanations of why even sound arguments with true premises often fail to prevail.

An account in terms of paradigms appears, at first sight anyway, to be able to accomplish all the requisite explanatory work. The different perspectives involved are competing paradigms: the "alternative environmental paradigm" as against the "dominant social paradigm", both discerned, with lesser or greater clarity, in much recent literature. These paradigms are suitably exhaustive. Moreover, this competing-paradigm picture can, with but little adjustment, explain why public projects, like forestry and electricity schemes and dams, proceed even if at a serious public loss and with a negative cost-benefit assessment under the prevailing paradigm. It is because – at a more superficial level – some private group with political influence stands to make a return, and because – less superficially – of a *commitment* (to put it mildly – "mania" would sometimes be a more appropriate term) to development and management within the prevailing paradigm which overrides the economic criteria the paradigm yeilds.[1] Observe, however, the seeds of destruction of the picture within such explanations. The realistic admission that elements of a paradigm may compete, and that conflict may be resolved by a ranking, already makes room for the hypothesis that two different, if overlapping, paradigms are operating, and thereby also raises the awkward question of identity conditions for paradigms. (As we all now know: no entity without identity.)

Such an attempt to set arguments and issues within the picture of competing paradigms appears to have much to recommend it. Firstly, it reflects the way environmentalists often see the issues themselves, and the way they find themselves shifting grounds drastically according to the different parties they are talking with or against. Often, for instance, they regard themselves as pitted against (filthy) capitalists who stand to make (immense) profits out of offensive environmental despoliation or, more recently, against bureaucracies committed to narrow economic and environmentally destructive developments. They then discover that the despoliation, as well as infringing environmental values, fails to pass conventional economic criteria: the despoliation can be condemned, not only on deeper environmental grounds, but even upon a resource conservation model. The picture also reflects the way those who adhere to the dominant paradigm sometimes assume that there are environmentalists who cannot be argued with because they refuse to admit rational assessment methods (i.e., those supplied by the dominant paradigm) or, a little differently, because they adopt crazy and politically unrealistic values. The rift exhibits, whichever way it is looked at, the expected features of paradigmatic estrangement.

Secondly, the paradigmatic picture offers the prospect of a general framework for environmental philosophy, in terms of which many apparently diverse problems and issues can be assembled and organized, and a unified account of them, as manifestations of *the one deeper phenomenon,* can be given.[2] For *many environmental and social problems* (e.g., those concerning indigenous peoples) *have a common underlying structure.*[7] Indeed, the paradigm picture holds out the prospect of unification in much the way that earlier pictures (such as those in terms of forms of understanding, forms of consciousness, forms of life) and conceptual schemes appear to offer syntheses.

It is a pity that the paradigm theory doesn't work well without many repairs, as efforts to present the picture in detail soon show. We encounter two main trouble spots. It is far from clear what the paradigms, or perspectives, look like in proper detail; and it is doubtful that the different sorts of usually very incomplete objects that have been passed off as paradigms ought to count as paradigms (e.g., Kuhn even says that textbooks can be scientific paradigms, and Daly tries to present the book he edited in 1973 as "part of an emerging paradigm shift in political economy";[3] but mostly the paradigms offered are very impressionistic objects with only a few features). It is worth some effort to try to remove these troubles and effect repairs, and not merely in order to reinstate the initial picture. For the question of what these paradigms, or perspectives, look like is in any case of much independent interest for many reasons. For example, it may assist in determining whether a given problem can indeed be solved under a different paradigm;[4] in trying to get to the bottom of the (ideological) differences separating environmentalists from developers and economists; in trying to discern the new paradigm to which a change of consciousness is supposed to transport us and what was in the old paradigm we left behind, and in trying to evaluate respective positions. It may also help in trying to reveal the range of the possible and the feasible to those locked into one of the paradigms, especially the dominant one;

and in trying to assess such theses as the factual-transition thesis, that "our culture is undergoing . . . a major paradigm shift, a major transition to a post-industrial culture" and the normative-transition thesis that "our culture . . . is in need of a major paradigm shift" or ought to be undergoing such a transition.[5] It is important furthermore to look at paradigms, if we can, not only to set down how things are ideologically and how they are changing, but also in order to change things in order to see in an organized way how and where to try to alter practices. For, like Marx, many environmental thinkers are not concerned merely with how the world is, or how to interpret it, but how to change it. Paradigms, as world views, can play a central (intellectual) role in such assessments.

But while it is no doubt interesting and important to determine certain paradigms, it is not easy to capture them in detail. In fact determining cultural paradigms (i.e., the paradigms pertaining to whole cultures and not merely to, or within, a discipline) is a difficult and demanding exercise in the history of ideas. There is moreover a preliminary problem: being sure what paradigms are. Otherwise, since what is being sought is obscure, the search for paradigms will be that much the harder, and it will likewise be harder to know whether the search has been successful. The matter of what paradigms are, and are like, is similarly of independent interest, for example, in the philosophy of science in the aftermath of the Kuhnian "revolution".

I
Problems with the Notion of Paradigm

As already hinted, "paradigm" has become something of a vogue term in the social sciences and associated reaches of philosophy. The term seems here to stay for the time being, though it will likely fade in the longer term in the way that one of its immediate predecessors, "conceptual scheme", did. For the present, "paradigm" has no real rivals, and wins by default as it were. None of the other terms in the general area will serve as well. Terms that would do, more or less, for the sweeping cases of cultural paradigms, such as "world view" and "ideology" will not do for narrower applications,[6] for example, within special sciences such as political science. None of the other likely candidates, "theory", "system", "position", "viewpoint", "perspective", "archtype", "myth", can really substitute or even fill all the places where "paradigm" now makes sense.

So it is singularly unfortunate that the term is on the road to ruination. It now signifies a multitude of different things, and it has picked up some bad philosophical associations (with erroneous fallibilism and rival-ways-of-seeing doctrines).[7] The multiple ambiguity and associations are due largely to Kuhn, who partly severed the term from its original sense of *example* or *pattern* and, while putting the term to new useful work, has overworked it and given it too many diverse roles — with the result, among others, that some of the fabric woven with it falls to dust as soon as it is seriously touched.

As to excessive roles first, Masterman claims to distinguish at least twenty-one different senses of the term in Kuhn's small book. While not all these are that distinct and *some* reduction might be effected, still (too) many differences would remain. An idea of the diversity can be obtained from the three main groups Masterman arranges the twenty-one senses into:

1. A *metaparadigm*[9] or *metaphysical* paradigm is, variously, a set of beliefs, a myth, a structure, a new way of seeing, an organizing principle governing perception itself, a map, a determinant of a large area of reality, a successful metaphysical speculation.
2. A *sociological paradigm* is a concrete, scientific achievement or one that is universally recognized; it is also likened to a set of political institutions or, differently, an accepted judicial decision.
3. An *artefact* or *construct paradigm* is an actual textbook or classic work, some instrumentation or a supplying of tools: it is also a grammatical paradigm, an analogy, a gestalt figure.

Quite a tangle. Nor are the groupings very satisfactory; for instance, a "successful metaphysical speculation" grouped under heading 1 is an "achievement" — which is given as falling under 2 — as is a "classic work", grouped under heading 3; "a map", put under 1, is not (except in an extended sense) a metaphysical object, but more like a construct and would be better located under 3 and 1; and so on.

Subsequent critics, including Masterman herself, have increased the size of the thicket. Although "Kuhn . . . never . . . equates 'paradigm' in any of its main senses, with 'scientific theory' ", so Masterman claims,[10] part of the *point* of paradigms being that they are pre-theoretical, critics of Kuhn associated with the London School of Economics have equated a paradigm with a basic theory or a dominant theory, and differently with a research programme. Others have equated paradigms with what were, with little doubt, their (terminological) antecedents; that is, conceptual schemes,[11] models and metaphors.[12] A glance at what is done by those who, like Drengson, claim to be employing "the Kuhnian notion of paradigm" or also, like Masterman, to be clarifying "the nature of a paradigm", will indicate how slippery the ground is in the paradigm thicket. Having identified a sociological paradigm as a (scientific) achievement, Masterman soon goes on to say that "seen sociologically ... a paradigm is a set of scientific habits" or just "a set of habits".[13] But to equate an achievement with a set of habits is just a categorizing mistake; for example, the latter has habits as members, but an achievement cannot significantly do so. A textbook can (almost) be an achievement, but not significantly a set of habits, etc.[14] This categorizing error originates with Kuhn, whose only implicit definition of a *paradigm* is as an achievement — one that has these two features: it is firstly "sufficiently unprecedented to attract an enduring group away from competing modes of scientific activity", and, secondly, "sufficiently open-ended to leave all sorts of problems for the redefined group of practioners to resolve".[15] This lax account lets through many examples that should be excluded, for example, kidnapping of batches of scientists, conversions or groups of scientists at, for instance, Billy Graham crusades.[16] Furthermore, it too narrowly excludes paradigms that do not emerge from scientific activity but, for example, out of philosophy or myth. In short, Kuhn's definition affords neither sufficient nor necessary conditions. What Kuhn goes on to say is nearer the mark than his definition; namely,

> that some accepted examples of actual scientific practices — examples which include law, theory, application and instrumentation together — provide models from which spring particular coherent traditions of scientific research.[17]

The key term here is *models. Paradigms are,* or provide, *models of certain types which carry research programmes as directives.* Then we can luxuriate, for a little while, in all the rich ambiguity of the term "model".[18] Many of the uses to which "paradigm" is put, uses which on other accounts are nonsignificant or unintelligible, become meaningful and intelligible, for example, talk of the "breakdown of a paradigm by the emergence of an anomaly within it deepening into crisis"[19] and of "new paradigms of human-environmental relationships".[20] Such talk does not make too much sense in terms of most of the equations Drengson offers — of paradigms with symbols, ideals, metaphors — or any of the philosophical accounts Masterman proposes, through achievements, artefacts, literal pictures, etc.[21] But it can all be made good, in one way or another, in terms of models in the modern logical sense, and one can be more specific about the structure of the models.[22]

To my disgust, it looked, for a time, as if this claim would have to be withdrawn, even before a beginning has been made on assessing its adequacy. For instead of being able, as at first naively hoped, to consult books on model theory and there locate a suitable definition of *model*, investigations came to a rapid dead-end. In a sample impressionistic survey I went through all the books specifically on model theory in the University of Victoria library (a small finite number) and encountered *not one* general definition. For the most part, what was offered was at best an account of models for first-order extensional languages, a *far* from adequate class for current purposes.[23] Fortunately, I then recalled some work of Routley which purports to give a universal model theory, good for any language.[24] In essence, what the universal theory does is to add to the sort of models adequate for second- and higher-order extensional languages a class of worlds, that is, it combines type-theoretic models, *reconstrued* as operating for free λ-categorial languages, with intensional (modal or relevant logic) models.[25] The general idea is, however, the same as in the more familiar first-order case; namely, the model (M) consists of both a *relational structure* or *system*(S) — which includes a class of worlds possible and not — and an *interpretation* or *valuation* (I on S), which *at least* verifies (i.e., makes true) or forces, a given (generally *any* given) class of statements of the language. In fact, a model can distinguish much more than a class of truths. It can also, when suitably characterized, yield the class of nonsignificant assertions, the class of assertions neither true nor false, the class of those that are both true and false, the class of rules endorsed, and so on. It can, in short, provide a full semantical theory and, when contextually enriched, the pragmatics as well.[26]

In terms of the general theory of models, recent philsoophies of science presented in terms of paradigms — more importantly, what seems right in them — can be re-expressed more exactly (or with the usual, commonly justified, logical illusion of precision). Of course, when so expressed, the recent philosophies cease to look quite so new or original, since model-theoretic accounts are a tiny bit older: but then originality and newness tend to be much exaggerated in philosophy. In fact, much of philosophy (that not adapted to advances in science) consists of not substantially more than terminological upgrading of older positions.

In the *refined* sense, a *paradigm* simply is a *model* of a certain type – which type has yet to be clarified)– just as certain American dictionaries say it is, but *without* the refinement, or clarification. So the account *does* have natural language roots.[27]

Given the account, much (not all) of the new talk about paradigms can be straightened out or made good, with of course some terminological bridges inserted. For example, a textbook or classic work is *not* a paradigm, but *supplies* a paradigm; an achievement is not usually a model, but what is achieved can likewise, if suitably propositional, supply a paradigm; etc.[28] In rather converse fashion, a paradigm can *give* a set of (scientific) practices – Masterman's "set of (scientific) habits" – through the class of methodological rules the model endorses. One of the distinguishing features of any model that serves as a (scientific) paradigm appears to be that it endorses a set of (methodological) rules and, thereby, delivers a set of (scientific) practices. It can simply be made a necessary condition for a model to be a paradigm that it endorse a non-null class of (methodological) rules.

But note that the adequacy of the ambiguous OED account of "model" to bring out all the required features of paradigms does *not* establish the adequacy of the refined sense for the same purposes, without further ado. To show the adequacy of the refined sense it needs to be established that all relevant uses of "paradigm" can be explained in this way. Only a promise that this can be done has strictly been given (since it is a considerable diversion from the main objectives). Note further, however, that the refined sense solves certain problems that the *Oxford English Dictionary* account leaves open, for example, the small matter of identity conditions. Paradigms are identical if their models are. Then too, in terms of underlying models, it is straightforward to define other needed actions; for example, there is a *paradigm shift* where one model is no longer adopted and a (significantly) different model is adopted (adoption being defined below).

In this (almost superficial) way, not only much of what Masterman has to say about paradigms in the philosophy of science, but also a goodly amount of what Lakatos says about research programmes can be absorbed.[29] For, according to Lakatos, a research programme consists of methodological rules, some of which tell us which paths of research to avoid (*negative heuristic*), and others, which paths to pursue (*positive heuristic*).[30] Then, a research programme, with respect to a given paradigm (P) is a (sub)set (MR) of the rules P endorses; while +H and -H are subsets of MT such that +H $\cap$ -H = Λ and the rules of +H and -H specify respectively research to be pursued and avoided.[31] The latter characterization may seem rather nebulous — it *is* rather nebulous — but it is all that Lakatos offers by way of characterization, though it is accompanied by some examples. However, both -H and +H have further important roles — some of which suggest (what will soon have to be introduced) (temporal) sequences of models. For example, -H determines a *hard core* HC among the statements not false on the model, namely those -H forces to nonfalsity. A (sub)class of statements outside HC constitutes the *protective belt* of auxiliary hypotheses which are not shielded from falsification. Of course, as in Lakatos, this is mostly suggestive terminology: tighter conditions on all these objects are really required.

To include other things Lakatos introduces, and leading features of Kuhn's theory, including his account of *normal* and abnormal science, it is necessary to consider sequences $P_1 : i \epsilon I$ of paradigms. According to Kuhn's theory the progress of science is represented by sequences of *operational* paradigms, that is, paradigms which are *accepted*, (i.e., have a sufficient support basis). Kuhn's other requirement on when "achievements" are paradigmatic, that they carry a research programme, is already satisfied by virtue of the definition of "paradigm". But pursuing this line leads us through the *theory of development*, a way followed elsewhere, but it is not our present route.

To include the social or cultural paradigms of interest in the social sciences and in environmental philosophy, some alteration and broadening of the class of models indicated by Kuhn and Lakatos, and of those suggested primarily by single discipline paradigms in the physical sciences, is essential. In particular, cultural paradigms do not carry research programmes, nor do they give rise to "particular coherent traditions of scientific research": such characteristic features of Kuhnian paradigms as carrying detailed research programmes are mostly entirely neglected by social scientists, who take over what they claim to be the Kuhnian notion of paradigm. However, more generously construed, paradigms may be said to carry *elaboration procedures*, which include defence strategies, etc.; and these can be taken to be given, once again, by methodological rules. Furthermore, what social scientists and philosophers usually want to include among paradigms are objects much more like what are sometimes called world views or whole-subject views,[32] ideologies or (general) conceptual schemes — all rather sweeping perspectives or outlooks, involving values and value assumptions, beliefs and practices, and influencing ways of perceiving issues.[33]   Perusal of what social scientists have to say about paradigms, and the senses of "paradigm", helps to confirm this. Not uncommonly, they distinguish among senses a "broad sense of paradigm", for example, along the lines Rodman characterizes a "cultural paradigm" or "dominant social paradigm", as consisting of the "basic beliefs, values, political ideals and institutional practices of a cultural epoch".[34] Often *paradigm,* in this broad sense, is equated with *world view.*[35] What distinguishes models (or interpreted systems) of this type is certain of the propositions or assumptions involved which concern the way the world is or is viewed and/or the way it should be — what can be called *world propositions.* Thus, for example, the first assumption of what Catton and Dunlap refer to as the "Dominant Western World-view", in contrast with the "New Ecological Paradigm",[36] is the following world proposition concerning humans: that people are fundamentally different from all other creatures in the world, and (rightly) have dominion over them (see table 4, p. 000).

The refined sense of *paradigm,* introduced as a type of model in the technical sense, can accommodate all this, with some elaboration. To indicate how, the class of rules the model endorses will include both *elaboration procedures* (which will lead to new models) and rules giving *practices.* Within the domain of objects the model has will be a subdomain of values, about which the model will determine *value assumptions,* on particular values held, through the interpretation function. A *world* paradigm will validate some world propositions. A creature (Z) *adheres to* a paradigm if Z accepts its values (i.e., assumptions as to values held) and believes its assumptions; in brief, if Z's revealed beliefs include its assumptions. And so on. Fanciful stuff, but with the *ring* of science and logic.

II
Dominant and alternative World Paradigms

There is now an abundance of literature about, or alluding to, paradigms or paradigm changes, which scarcely addresses the question of what change is being suggested or what goes into the prevailing or the new paradigms.[37] Even where some details are set down[38] — invariably in shorthand and often virtually in note form — there is much gesturing as to how things are to be filled out, if they are at all. This makes testing of the claims made, assessment of theses about paradigm transition, for example, difficult or impossible. To know whether and what paradigms are adhered to, and whether they are shifting, it is generally necessary to know what these paradigms involve in some detail, how they are shifting, etc. Otherwise it is like asking about the foundations of a social science (or even a natural science): one is on shifting sand.

Fortunately for the present derivative enterprise the occasional philosopher, notably Drengson, has tried to articulate dominant and alternative paradigms, and the supposed contrast has been tabulated by certain sociologists (who are less bothered about looking crude or silly than philosophers). Those sociologists who have felt obliged to test hypotheses, have been — have been forced to be — more specific and precise (and thereby sometimes more simple minded) about these things than philosophers and political scientists; and we can begin at the more exact end of recent work with some of their tables, which are usually presented together with a small but very incomplete commentary. It is widely assumed, though the evidence goes against this, that there are just two rather monolithic world paradigms, the old or dominant and the new or alternative (sometimes both are said to be [co] dominant). The intended contrast is shown in synoptic form in table 1.

As you can readily see, this highly abbreviated presentation[39] makes many questionable assumptions, *some* of which will be brought out in what follows. To begin, the table, which Cotgrove and Duff present in this bald form, leaves *much unexplained,* for example, the last two pairs of contrasts are not explained at all, though their import is not completely evident. Thus it is quite unclear whether the integrations suggested in the final item are a necessary part of *an* alternative environmental paradigm at all.

The competing paradigms are in fact *far from exhaustive* of contemporary paradigms, in several respects. One, which is of crucial importance for the arguments sketched at the outset, is that main, present-day economic alternatives to the neoclassical economic position, which the items listed under "Economy" crudely represent,[40] are not included among the two alternatives. But it would hardly suffice to argue that because large-scale nuclear development is justified neither under environmental alternatives nor under neoclassical assumptions, that it is in no way or nowhere justified, without at least considering other major ideological positions under which such development is often claimed to be justified, namey, state socialism with its

Table 1. Attempt 1 Competing Social Paradigms

	Dominant Social Paradigm	Alternative Environmental Paradigm

Core values	Material (economic growth)
	Natural environment
	  valued as a resource
	Domination over nature
		Nonmaterial (self-actualization)
		Natural environment
		  intrinsically valued
		Harmony with nature

Economy	Market forces Risk and reward Rewards for achievement Differentials Individual self-help	Public interest Safety Incomes related to need Egalitarian Collective/social provision

Polity	Authoritative structures
	  (experts influential)
	Hierarchical
	Law and order
		Participative structures
		  (citizen/worker involvement)
		Nonhierarchical
		Liberation

Society	Centralized Large-scale Associational Ordered	Decentralized Small-scale Communal Flexible
Nature	Ample reserves Nature hostile/neutral Environment controllable	Earth's resources limited Nature benign Nature delicately balanced

Knowledge	Confidence in science
	  and technology
	Rationality of means
	Separation of fact/value,
	  thought/feeling
		Limits to science
		
		Rationality of ends
		Integration of fact/value,
		  thought/feeling

NOTE: Reprinted, by permission of the publisher, from S. Cotgrove and A. Duff, "Environmentalism, Values and Social Change", *British Journal of Sociology* 32, no. 1 (1980):341. This modest table (slightly altered here) is the most comprehensive I have encountered in the literature.

"command" not "market" economy and democratic socialism with its mixed market-nationalized economy. Upon setting out these economic variants upon the *neoclassical* (or American) social paradigm" (the "mainstream paradigm" in Samuelson's terms, no doubt, since it reflects "mainstream economics"), *the* dominant social paradigm splits into three impure social paradigms, the other two being the (official) *state-socialist* (or Eastern European) paradigm" and the *social-*

democratic (or Western European) paradigm", these latter paradigms differing primarily as regards items listed under "Economy" (e.g., *profit* is replaced by *surplus value*). While the listings "Rewards for achievement" and "Differentials" stay put (as they would not under communism), other items listed do not.

It is by now evident that we can readily reach a quite embarrassing position for those who suppose that paradigms exist but like to keep their ontologies clean. For not only can we go up, to an over-arching dominant world paradigm, which drops out continental economic differences, but retains the large area of overlap (under headings other than "Economy"; but we can proceed down to regional, or even national, paradigms, for example, the French as opposed to the Scandinavian. There is a reasonable basis for such regional talk; it is common for authors to refer, for instance, to the Yugoslav model of self-management. Rodman, in his discussion of political paradigms, chooses to go up from differences between market and command economies in his sketch of the "modern" paradigm.[41]

On the one hand, however, it is often historically important (as Rodman is aware) to include factors of economic significance — for instance, property and interest as well as (certain) markets are crucial elements under the Enlightenment paradigm of the eighteenth century (i.e., in effect, classical economic theory has furnished the economic assumptions of that paradigm) — if the paradigm is to have an appropriate explanatory role (in a history-of-ideas setting). What Rodman manages to reflect (see table 2, p. 000) is not *the* modern paradigm, which has evolved in a sequence from Renaissance through Enlightenment to colonial capitalism to its contemporary successors, but an amalgam consisting primarily of a common denominator of the modern sequence, that is, certain common elements of these evolving paradigms with, however, lapses to elements which are not common (e.g., under headings 2 and 3 in table 2).

On the other hand, for some arguments (e.g., as regards New Zealand forestry) it is important to consider quite local markets. So *which* are the paradigms: the dominant world paradigm, or the three continental paradigms,[43] or the models that fall under them, or . . . ? The answer, at least from the point of view of object theory, is easy: all are, so long as they satisfy the characterization of *paradigm*. But the lower models will not count as world paradigms if the propositions they verify are regionally restricted.

Paradigms have, to an interesting extent, a tree structure, as the dominant case reveals:

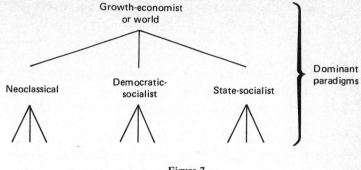

Figure 7

Discipline or subject paradigms (of the sort on which Kuhn focuses) also fit in, at each time, to a tree structure, of the following form:

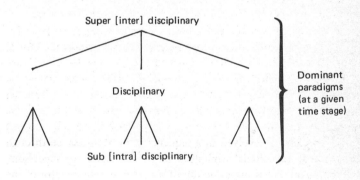

Figure 8

Such paradigm trees will of course graft onto the cultural-paradigm tree. The structure of paradigm interrelations is still more complex, because at many nodes in dominant-paradigm trees there are rival paradigms to the dominant paradigms, and also because sometimes there are various competing paradigms with none dominant. The latter appears to be the situation in sociology and political science, where various "paradigms" (structural-functional, etc.) vie for dominance. The more complex picture can be envisaged as that of a *paradigm forest* with comparable tree nodes interlinked with vines; and in the forest will occur some dominant trees (with the structure diagrammed above, but upside down).

Do rival or alternative paradigms also decompose into constellations with tree structures? They certainly decompose, but into many intertwined trees. For *there is not,* it seems, *a single alternative environmental paradigm,* but, to be more accurate, various different *suggestions* for alternatives to the dominant paradigms. The modelling of cultures or ideologies in terms of paradigms tends to give a misleading impression of monolithic and uniform positions, like the two paradigms that appear in several presentations of the contemporary ecological predicament, or the one paradigm that covers each past period of historical interest, such as the (dominant) paradigms lightly sketched[44] for classical antiquity, for Renaissance Europe, and for Enlightenment Europe. But even in turbulent intellectual times, such as fourth-century B.C. Greece, there appear to have been competing paradigms and perhaps paradigm shifts occurring; so the picture presented is deceptively simple. Rodman's contrast of a single classical paradigm with a single modern paradigm (see table 2, p. 000) illustrates this point (and others).

There were, however, rival paradigms with some support base in classical times as in contemporary times. For example, there appears to have been a Greek counter-culture, in the form of the Cynic movement, with very different values, assumptions and practices from those of the dominant position now portrayed to us primarily through Plato and Aristotle. The problem this makes for Rodman is, however, easily avoided, either by adding the qualification *dominant* (bracketed in the heading above) or else doing what often appears to be done (by Kuhn included), equating paradigm with dominant paradigm. (Then other paradigms become, say, paradigmatic models).

The contemporary alternative environmental paradigm fragments to such an extent under examination that there is room for genuine doubt as to whether there are alternative models with requisite support to count as *operative* paradigms. Of course there are alternative positions and models. Reaction to the contemporary environmental predicament has led to a wide range of proposals and positions, ranging from far-left-leaning anarchist positions (e.g., Bookchin), to far-right-leaning state-authoritarian positions (e.g., Hardin, Heilbroner). The range also includes positions (like Rothbard's market- and property-based anarchism) which profess to be neither right-leaning nor left-leaning and which may contain a genuine mixture of leftist and rightist

Table 2. Attempt 2 Contrast of a classical paradigm with a modern paradigm

	(Dominant) Classical Paradigm	(Dominant) Modern Paradigm

Ideals	1. Virtue and its aspects:
	    wisdom
	    justice
	    temperance
	
	but NOT:
	    wealth
	    power
	    sensual pleasures
		1. Mercantile virtues:
		    wealth
		    power
		    possession of property
		      (of declining import)

	2. Self-sufficiency for:
	    individual
	    polis
		2. Scope for individual:
		    liberty and freedom
		      (with qualifications)
		    toleration (within
		      limits)

	3. Emphasis on relatively "spiritual" lifestyle	3. Human (self-)realization, in part through domination and transformation of nature
Exemplar	Sage and monk	Entrepreneur and technocrat

Further themes	1. Limits on
	    City size
	    Wealth
	    Poverty
	    Individual freedom
		1. No limits: freedom as progressive liberation of Man from traditional and natural limits *

	2. Old Differential Imperative (i.e., maximization of species-specific differences), achieved by humans transcending their animal and vegetative nature, and by the domestication of wilderness	2. New Differential Imperative, achieved by human domination of nature, and monocultural reduction
	3. Cosmic control of universe: teleological explanations often required	3. Mechanistic explanations only
	4. Doctrine of mean: virtues as the mean between extremes Justice as: keeping one's place and performing one's function Giving everything its due	4. Economic-technocratic rationality 5. Nature a natural resource with no intrinsic value 6. (Other) Animals as machines

NOTE: After John Rodman, "Paradigm Change in Political Science", *American Behavioural Scientist* 24, no. 1 (1980): 51ff.

* The no-limits theme has to be handled with some caution (more than Ophuls, e.g., observes). For neoclassical economic theory is *based* upon a scarcity assumption. While scarcity and limits can, to a sufficient extent, be avoided by *substitution* (which Ophuls scarcely considers), scarcity can never be removed because (so it is mistakenly assumed) human "demands" are unlimited.

elements, and, of course, inconsistent positions (such as Ophuls' appears to be) which try to combine, for instance, centralist governmental control with decentralist nongovernmental freedom. Because of the variety of models and positions indicated in the literature, the so-called alternative environmental paradigm, compiled (as the sociologists' table appears to be) by listing opposites (of a sort) for each element of the neoclassical paradigm, falls apart. Alternative paradigms are far from uniquely determined: there is a variety of alternatives most of which are not operative, but some of which may be. Consider, for example, the resource-conservation position based on long-term human interests. While the position is outside the dominant paradigms and shares some features with Cotgrove and Duff's alternative, it diverges fundamentally from it, for instance, as to whether the natural environment is intrinsically valued. The particular steady-state economy Daly has presented in several places as "an emerging paradigm", and which has some support, especially among scientists, could serve as an illustration, since it is left unclear what else but steady-state conditions go into this incomplete "emerging paradigm". In principle, Daly's "paradigm" need differ from the neoclassical paradigm only as to growth and reserves assumptions,[45] and so would afford only a rather shallow alternative. Likewise, diverging from *the* alternative, and sometimes committed only to a shallow alternative, are many right-leaning environmentalists who (persuaded by the "tragedy of the commons" or similar arguments of a prisoner's dilemma cast) believe that solutions lie through powerful central governments with tight controls incompatible with participative structures and egalitarian practices. The alternative proposed comprises, then, a special package which would not be accepted in toto by many critical environmentalists. (It would appeal to some of those favouring a form of small-scale communism.)

It *may* begin to look, then, as if there is but a flimsy case for discernment of the alternative paradigm. Certainly there appear to be interesting, theoretically coherent positions of one sort or another, one presumably corresponding to the alternative, but do any of them command a sufficient support base to be operative as paradigms? Evidence that they do has not so far been assembled, it seems. The data that has been assembled shows attitudinal variations, along several dimensions, from themes of dominant paradigms, as indicated in table 3, p. 000, which reflects Cotgrove and Duff's evidential base much more accurately than does table 1.

Had an operative alternative paradigm been *empirically* discerned,

Table 3 Attitudinal variations from dominant paradigms

A. Wealth creation		Limits to growth
Economic growth	o+* Δ	No economic growth
Unselective production	o *+Δ	Selective production
No limits to growth	o *+ Δ	Limits to growth

B. Authority		Participation
Management	o * Δ +	More say at work
Needs of industry	o* +Δ	Satisfying work
Government	o *+Δ	Participation
Experts	o * +Δ	Participation

C. Market		Non-market
More law and order	o * +Δ	Less law and order
Achievement	o * Δ +	Other criteria
Market	o *Δ +	Non-market
Differentials	o *+ Δ	Egalitarian

D. Individual		Collective
Individual	o * Δ +	Community
Individual welfare	o *Δ +	Social welfare
Individualism	o *Δ +	Community

KEY: Public (at large) *   Industrialists o   Trade Unionists +
Environmentalists Δ

NOTE: Reprinted, with changes, by permission of the publisher, from Cotgrove and Duff, "Environmentalism", p. 99. The table is somewhat confused and confusing: e.g., it is unclear how individuals and individualism, *both* contrasted in the community, are supposed to be distinguished; under Market is included not only market but broader matters, as well as questions of polity, etc.

testing would have revealed requisite clustering at least by some sufficiently prominent group about the distinguishing features of the paradigm. In the case of an environmental paradigm, it would not be unreasonable to suppose that environmentalists, or an independently isolated subgroup of them, agreed as regards features of the paradigm. Table 3 shows that for environmentalists as a group that is not so: the environmentalists they sampled are biased towards more law and order and are achievement oriented, for example. The required clustered divergence from dominant-paradigm themes remains to be

Table 4. Attempt 3. A comparison of major assumptions in the dominant Western world view, sociology's human-exemptionalism paradigm, and the proposed new ecological Paradigm

	Dominant Western World view (DWW)	Human-Exemptionalism Paradigm (HEP)	New Ecological Paradigm (NEP)
1. Assumptions about the nature of human beings	People are fundamentally different from all other creatures on Earth, over which they have dominion.	Humans have a cultural heritage in addition to (and distinct from) their genetic inheritance, and thus are quite unlike all other animal species.	While humans have exceptional characteristics (culture, technology, etc.), they remain one among many species that are interdependently involved in the global ecosystem.
2. Assumptions about social causation	People are masters of their destiny: they can choose their goals and learn to do whatever is necessary to achieve them.	Social and cultural factors (including technology) are the major determinants of human affairs.	Human affairs are influenced not only by social and cultural factors, but also by intricate linkages of cause, effect and feedback in the web of nature; thus purposive human actions have many unintended consequences.
3. Assumptions about the context of human society	The world is vast, and thus provides unlimited opportunities for humans.	Social and cultural environments are the crucial context for human affairs, and the biophysical environment is largely irrelevant.	Humans live in and dependent upon a finite biophysical environment which imposes potent physical and biological restraints on human affairs.
4. Assumptions about constraints on human society	The history of humanity is one of progress; for every problem there is a solution, and thus progress need never cease.	Culture is cumulative; thus technological and social progress can continue indefinitely, making all social problems ultimately soluble.	Although the inventiveness of humans and the powers derived therefrom may seem for a while to extend carrying capacity limits, ecological laws cannot be repealed.

NOTE: Reprinted, with changes, by permission of the publishers, from W. R. Catton Jnr. and R. E. Dunlop, "New Ecological Paradigm for Post-exuberant Sociology", *American Behavioural Scientist* 24, no. 1 (1980):36.

Table 5. Attempt 4.

(Cartesian) Technocratic Paradigm	(Organic) Person-Planetary Paradigm
Atomistic analysis; discrete elements	Organisms as wholes; fields and processes
Mechanistic reduction	Symbiosis, mutual interrelations; consciousness irreducible
Separation of subject-object, science vs. spiritual	Integration and reciprocity of subject-object, science and spiritual
Material growth (favouring high technologies)	Homeostasis and balanced development
World as artefact: nature only a resource, to be mastered	World as living organism (Gaia hypothesis)
Anthropocentricism: value in nature instrumental	Biospheric egalitarianism: intrinsic value in being itself
Efficiency and productivity; design as technique	Ecological design of human activities; design as art
Modern technology: repetitiveness, uniformity, predictability, interchangeable parts, specialized skills: technological knowledge as power	Appropriate technology: diversity, open possibilities, spontaneity with order, balanced education: no power play but understanding and spiritual dimensions
Government by expert-underpinned elite; corporate control	Collective responsibility; community based
Structure monolithic, centralized, manipulative, capital-intensive, labour-poor, competitive; individualistic	Decentralised, nonmanipulative, egalitarian, cooperative; social
Science value-free and objective	Human experience value-laden; personal knowing
Formalization and quantification methods (linear, one dimensional emphasis)	Dialectic and quantitative procedures (multidimensional emphasis)
Every problem soluble, through technique, and specialization	Limits; cultivation of whole person
Determinism: law and principles of order	Freedom in community
Persons evaluated in terms of their roles	Unique value of individuals
Persons as mechanical, closed, in need of control, isolated	Persons creative, open, developing; intersubjective experience and diverse consciousness
Personal ideals: wealth, power, influence	Personal ideals: wisdom, understanding, self-mastery, openness
Operative personal stance: *having*	Operative stance: *being* (and *doing*?)

NOTE: After Alan R. Drengson, "Shifting Paradigms: From the Technocratic to the Person-Planetary", *Environmental Ethics* 2, no. 3 (1980):221–40. The revised compilation of this table takes advantage also of a schematic comparison of the two paradigms (a philosophy-class handout) supplied by Drengson. Some liberties have been taken in revision (e.g., neglected contrasts brought out, others improved).

Still again there is *much* to criticize in this value-loaded table. Some of the more serious weaknesses, often shared with other tables, are suggested in the text. More generally, too much of the paradigmatic contrast reduces to that between Cartesianism and dialectical nonmaterialism, both of which are too specialized.

shown, especially as several of the issues concerned are — at least super-ficially — independent of one another, unless unified through a deeper, underlying viewpoint.

Of course some convergence — by a small group like "small-scale-communistic" environmentalists — would be expected, though this has not been confirmed. But unanimity from such a *sub*class of environ-mentalists (themselves commonly comprising less than five per cent of a population) would hardly support the triumphal emergence of an alternative paradigm, that is, *emergence* in the sense that there is a *substantial* operational base. Such empirical questions are not without philosophical interest, since operational environmental paradigms help remove the familiar "straw-people" charge from environmental philosophy and assist in legitimizing it as a subject in administrative eyes.

Another decidedly awkward aspect of claims about the dominant social paradigm is that (like those about the alternative) every author who bothers to set down many of its features provides a different bundle of features. Compare the attempts to articulate the paradigms already tabulated with the further attempts in table 4, p. 000, and table 5, p. 000.

It is evident from inspection that the attempts presented, and also less detailed accounts elsewhere, have a great deal in common. The situation can be depicted by a familiar Venn diagram:

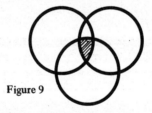

Figure 9

Some of the differences between the attempts can be rectified to *some extent* by additions, as will be seen. Moreover, some of the regions where the attempts differ can be removed, as they are based on mistakes, for example, Drengson's attempt to include in the alternative paradigm such themes as (insufficiently qualified) biological egalitar-ianism and the Gaia hypothesis.[46] Sometimes the attempts are really directed at isolating paradigms: for example, Drengson, in discerning a Cartesian "technocratic" paradigm, can be better seen as trying to isolate a *sub*paradigm of the world paradigm, namely one that conforms to a Cartesian dualistic-cum-mechanistic philosophy, rather than one which abstracts from types of philosophical positions, and isolates what is common to those falling under dominant paradigms (e.g., what is common to empiricism, pragmatism and Cartesianism).

The procedure illustrates an heuristic method for arriving at higher paradigms, a method already applied in working towards the "growth-economistic" paradigm, namely, that of removing elements which are different in certain locally competing paradigms to obtain consensus, for example, removing the continental differences between the "dominant growth" paradigms. While this amounts to working up the tree, there is some art (or perception) required in deciding which items to vary and which to hold constant. Higher environmental paradigms can be reached by the same method.

The incompleteness of earlier tabulations, is bothersome. For instance, what have been omitted from table 1 are not only contrasts under the headings supplies, such as, under "Polity," the contrast between institutionally channelled action and direct action; but, more critically (as Drengson's discussion indicates), a major group of contrasts — "reductionist/nonreductionist" and "analytic/synthetic, holistic" — under what might be headed "(Philosophical) method".[47] Methodological rules might have been expected to deliver such contrasts early on. Thus, mere deletion is insufficient in reaching high dominant paradigms: even worse, addition is also required.

To arrive at high paradigms, then, previous attempts such as Cotgrove and Duff's have to be elaborated, as follows in the case of the dominant paradigm:

1. Deletions are required, especially under the heading "Economy".
2. Additions are called for, especially under a new heading, "Method".
3. The tables need to be expanded into fuller form, and also directions given as to how the fuller forms could be presented as paradigms in the technical sense.

Under such a high dominant paradigm, other paradigms such as neo-classical and state socialist would fall. Reaching high environmental paradigms is a still more complex, and hazardous, affair. It looks as if a tree structure for environmental paradigms should tie in with a classification of environmental positions. Thus, a high paradigm would include qualified or selective economic growth and limited-resources themes, but leave open (and thus be deliberately incomplete, as regards) the crucial matter of intrinsic values. And under this high paradigm would fall shallower and deeper environmental alternatives.

Even when done, the results would inevitably be incomplete, and exhibit a certain arbitrariness and ideological bias. Neither of these features undermines the project of making a synthesis of attempts to discern paradigms. Nor is it unrealistic to try to set down with some exactitude what the high cultural paradigms are like. And there are good reasons for trying, in terms of eventual philosophical returns. As to why it's not entirely realistic but why it is worth trying nevertheless, compare the business — important for verification or refutation — of determining philosophical positions, especially commanding positions which dominate the philosophy of a given time. Nor is the situation quite as bad as it may appear to be. For many purposes it is enough to know necessary conditions, or perhaps sufficient conditions, for given paradigms.

Incompleteness is to be expected, and is rather inevitable (so it is perhaps less damagingly described in terms of *openness*). It is to be expected because these paradigms are intended to reflect creatures' beliefs as to certain notions and values, and creatures' beliefs are characteristically and (almost) invariably incomplete. Indeed the situation is typically worse than that: "Converse suggests that only about 10 per cent of the population have complex 'belief systems' that deserve the name of ideology, and it is only among these usually rather well-educated or involved people that theories of political ideology and motivation really hang together".[48] Thus, unless models remain incomplete (and *without* depth of connection between assumptions encompassed) and exhibited in seriously incomplete form, they will lose contact with their empirical base.

Incompleteness is rather inevitable because there are infinitely many values it seems, but only assumptions concerning finitely many are included in the relevant paradigms. In fact, the class of values considered is quite restricted, and it might well be asked[49] why such common values as good health and personal survival and security and eating well are not included in any of the tables. An initial answer is that the paradigms concerned are socio-environmental paradigms which do not involve what are classed as "private", or personal values, but only more "public", or political, values. This distinction has been said to be needed to resolve an apparent paradox in the attitudes of those accounted postmaterialists,[50] and can be discriminated sufficiently sharply for sociological-testing purposes (by linking "private" values with life-satisfaction questions). The values included in the paradigms given coincide in type exactly with those listed by Marsh under *political values* and are discrete from those listed under *personal values*.[51]

While this classifies what is included in the paradigms, it does not really explain what is included. What have been produced as social, or cultural, paradigms are concerned only with political-economy values (such as would be considered under political economy broadly construed): they are not full cultural paradigms, but are embedded in more comprehensive cultural paradigms which also take account of personal values and their rankings.

III
Paradigm-shift and transition theses

Since there is evidence of attitudinal shifts away from dominant paradigms along several dimensions,[52] a straightforward (weak) form of the factual-transition thesis holds. Nor is the shift a merely attitudinal one which has no bearing on practice – it has a considerable bearing on preparedness to protest – even though, too often, changes in attitudes (as expressed) are not accompanied by corresponding changes in practices. Inglehart's thesis is that there has been a substantial change in values held in Western Europe since 1945. Other sociologists claim to have substantiated this thesis,[53] and to have shown that some of the changes concern environmental and political values. Thus, for instance, Marsh claims that

> strong support exists for Inglehart's basic thesis. Political change in Western Europe, characterized especially by the growth of new non-class issues concerning participation, liberation, anti-industrialism, and the growth of unorthodox political behaviour has been fueled by the growth of new value orientations.[54]

However, sociological evidence so far to hand appears to confirm only a weak transition thesis, which is much more limited than the strong thesis Drengson presents (see n. 5). In fact, Drengson presents no evidence for the stronger factual thesis,[55] and it can be reasonably asked whether the evidence is there to be collected. If there were such a major shift going on, one would expect much more evidence of it in practice – whereas superficial observation tends to suggest that for most people in most industrial places little has changed: that for the dominant paradigms it is business almost as usual. However, in these areas attitudes are difficult to translate into practice: and there does appear to have been a *substantial* alteration in attitudes in recent times, alterations that the sociologists are beginning to record,[56] such as "the silent revolution", and that are increasingly affecting company and organizational practices (though mainly at the public-relations and environmental-impact-statement levels so far). The documented changes are not confined to environmentalists who form only a smallish fraction of the population, but extend to the general public. (It will require a *much* greater shift to build up to avalanche conditions.)

Anyone who adheres to an alternative paradigm that differs in major respects from the dominant paradigms will assent to the normative transition thesis. The thesis is contextually related to a given value system, and contextually analytic on the adoption of a suitable system. So the argument for the thesis boils down to the case for adopting an alternative paradigm. And the types of argument for that are, at least in outline, well known.[57]

What ought to be is not what is: we are far removed from the ideal worlds of alternative paradigms (i.e., worlds where they are realized). How are we to move towards the practical adoption of such paradigms? What action is to be taken to shift prevailing thought in deeper environmental directions and to counter the translation into practice of dominant social paradigms? Before (rather familiar) answers to such questions are outlined, there is a crucial complication to be considered. The problem arises that there is no necessary connection between practices and ideals, with the result that seeking a change in paradigms should not be the first or only objective of environmental action against the intellectual foundations of contemporary industrial society. This is especially so since an important route to changing beliefs and attitudes is *through* changing behaviour and practices.[58]

IV
Defects of the Attitudinal Approaches

Briefly, practices that are followed diverge, often sharply, from the position or attitudes that responses to questions indicate. This happens both with regard to the dominant paradigms and more environmental alternatives. Consider the more important *dominant case*.[59] Practices are increasingly distorted from neoclassical (or mainstream) ideals and assumptions by the role of large organizations (including governments, multinational corporations, etc.). Under the present arrangements, of what is sometimes called *advanced corporate capitalism*,[60] leading features of the usually discerned dominant-paradigm atrophy. In particular, the market diminishes in importance (e.g., it is displaced by such things as pre-ordering through subsidiaries or associates), and ceases to be "free" (i.e., does not approximate perfect competition at all) because of monopolistic or oligopolistic arrangements of large organizations, producer-manipulated demand, etc. Likewise, many other features applauded under the contemporary neoclassical paradigm disappear or are replaced: the individual (still the individual) who is most handsomely rewarded under the newer arrangements is not the entrepreneur but the organizational person — working for the organization is much favoured over individual self-help (sometimes it is pretended that these coincide), and so on.

A major problem intrudes here as regards the determination of paradigms, a problem that applies also to world views and ideologies, and indeed, in a more familiar way, to beliefs. Are we, in deciding what paradigm a person adheres to — or for that matter in defining paradigms — to take into account what a person says or what a person does? Do we put an industrialist solidly in the neoclassical paradigm because he says, of course, that he believes in market mechanisms, etc., or do we push him out because in his daily practices, his *revealed beliefs*, so to say, he regularly shuns market devices? The problem is not new, and is familiar with regard to religious beliefs and paradigms, in such shapes as "Sunday believers". We can distinguish an *espoused* paradigm and a *revealed* paradigm. It seems plain that if we want to know what paradigm a person *really* adheres to — like what beliefs he really holds — then it is the latter that matters. So paradigms can still be linked with beliefs in the expected way, that is an adherent of a paradigm believes in its themes or assumptions, and world paradigms retain their linkage with world views and ideologies.

However, questionnaires *may* not be a reliable guide to (revealed) paradigms as opposed to espoused paradigms, especially if the questions are naive — as they appear to be on the sociological testing that has so far been done. For example, in Cotgrove and Duff's survey, respondents were asked to indicate their preferences as to the sort of society they would like to see with questions like this: "A predominantly capitalist society, in which market forces and private interests predominate, or a predominantly socialist society, in which public interests and a controlled market predominate?"[61] To determine an industrialist's beliefs as distinct from his expressed views,[62] much more indirect questioning of types well known to sociologists would be required (since attitudes to the market are rather like those to motherhood and like those to God used to be), for example: "If you were buying Z's would you buy from an associate who supplies them or would you first shop about?" (There are doubts, too, properly raised about such formulations.)

Organizational objectives and goals do not coincide with those of neoclassical theory. While characteristically organizations tend to operate in their own interests,[63] their interests are not just in profit maximization — so they are *not* neoclassical firms — but include such objectives as maintenance or growth of the organization (self-preservation is commonly achieved through growth), removal of small "competitors", prestige, etc. (Galbraith and others have set out these features.)[64] Those who manage and work in large organizations, while they may frequently espouse a neoclassical paradigm, or something like it, often do not adhere to such a paradigm.

But it is *not* widespread adoption of the neoclassical paradigm that is tied with some of the worst of present environmental problems (unfortunately, because neoclassically based projects are often easier to argue against); it is the dominance of organizations and their practices that are at the source of many problems. Thus it is also important to delineate, so far as can be done, features of the organizational paradigms and their interrelation with dominant paradigms already indicated. There are several reasons why filling out the picture is important — for seeing what environmentalists are up against; for better estimating the prospects for change, regarding the questions of how to achieve change, that is, where leverage is pointful, where policy might be altered, how organizations might be modified, etc. Also important, and connected, are questions as to how organizational neofeudalism, in particular authoritarian trends within organizations and their impact on social life, can be reversed and how further drift toward totalitarianism avoided.[65]

Setting out a central part of the (obvious) organizational paradigm is straightforward since it is the same as the dominant world paradigm already adduced. Take the neoclassical paradigm of table 1, delete the "market" and "individual self-help" elements, and much of the organizational paradigm is apparent. Some entries should be modified, for example, the rewards are given for intra-organizational loyalty and achievement. In place of "profit maximization", which the neoclassical market presupposes, "maximization of organizational interests" appears. The organizational paradigm is a submodel of the world paradigm, and is opposed by at least those paradigms that oppose the world paradigm.

The divergence of revealed paradigms from espoused paradigms, especially of the exposed organizational paradigm from the espoused dominant neoclassical paradigm, complicates not only any theory of environmental thought which takes due account of what environmental positions are opposed to, but also the question of appropriate environmental action. For example, focusing on the neoclassical paradigm and directing polemics primarily against it would be, to some extent, misguided. Indeed, often enough, something would be accomplished by having components of the neoclassical paradigm genuinely adopted, such as in decision and policy making, in place of corresponding assumptions of the organizational paradigm.

What then is the role of alternative paradigms in environmental thinking and action? Requisite changes in environmental practices and attitudes will not be achieved by waiting for alternative paradigms to be adopted: indeed they are so far from being politically acceptable as to be commonly dismissed (as noted) as "politically unrealistic".[66] While alternative paradigms can fulfil all the roles of ideal models, for instance, providing bases for argument, positions to fall back on and around which to consolidate, and states to aim for, their main function is ideological; they are not where the main focus of action should be if requisite change is to be effected. That action has to be directed where and as it has been directed in the past — against the sources of environmental despoliation, primarily organizations; and those supposed to regulate them, further public organizations in the shape of governmental bodies — with a view to persuading them, or helping to make it so,[67] that the despoliation is not in the general interest, or in narrow organizational interests. Alternatively, the action may be directed with a view to blocking, delaying or otherwise hindering their despoiling operations, by both institutional procedures and direct action. If changes are to be effected there is no alternative — if *far too much* of the environment that is valuable is not to be lost[68] — to action (of a *broadly* political type). Philosophy, though relevant, is not enough.[69]

Notes

1. For example, the wood production ideology of foresters, discussed in Richard and Val Routley, *The Fight for the Forests*, 3rd ed. (Canberra: Research School of Social Sciences, Australian National University, 1975). Compare too, Rodman's suggestion that the

dam-building mode of "water resource development" has taken on the character of a ritual elimination of nature that is unintelligible in terms of conventional economic rationality. Dams are, after all, massive monuments to man's technological ability to tame the wild flow of ["waste"] natural energy and to substitute it into socially useful functions (irrigation, hydro-power, flood control, and the tamer forms of recreation). . . .

(John Rodman, "Paradigm Change in Political Science", *American Behavioural Scientist* 24 no. 1 [1980]:62.)
Of course in the full real-life picture there are other dynamic factors as well, (e.g., projects may have acquired momentum while having appeared to be economically satisfactory or while exponents of the projects have managed to project the illusion of satisfactoriness).

2. It would also help thereby in legitimating environmental philosophy. The competing paradigm picture has been latched onto by social scientists as a way of organizing discussion of an alternative intradisciplinary paradigm they want to promote: thus, H.E. Daly, introduction to *Towards a Steady State Economy*, ed. H.E. Daly (San Francisco: W.H. Freeman, 1973), pp. 1–6 in support of a steady state economy. Also see, William Ophuls, *Ecology and the Politics of Scarcity* (San Francisco: W.H. Freeman, 1977), said to be alternative political science. These organizational efforts likewise fail, as will emerge.

3. Daly, *Steady-State Economy*, p. 1.

4. Typically this will depend both on how the paradigm is defined and how the problem is specified, but it should not be *too* sensitive to specification. (Cf. Rodman, "Paradigm Change", pp. 49–78.)

5. Alan R. Drengson, "Shifting Paradigms: from the Technocratic to the Person-Planetary", *Environmental Ethics* 2, no. 3 (1980):221–40.

6. In addition, "world view" tends to have the wrong senses, when it means "view of life" or "contemplation of the world", the meanings given in the *Oxford English Dictionary*; and "ideology", which has been corrupted by Marxists, literally means "the science, or study, of ideas" and only by extension means – what is closer to, but still not at, what is often intended by "paradigm" – system of ideas.

7. More popular uses can be bizarre, for example, a "region on the brink of a new paradigm", and "paradigm" said to mean "a new set of circumstances" (*Evolution, Canada* 1, no. 1 [1981]:3).

8. Margaret Masterman, "The Nature of a Paradigm", in *Criticism and the Growth of Knowledge*, ed. Imre Lakatos and Alan Musgrave (Cambridge: Cambridge University Press, 1970), p. 65.

9. The abbreviation is unfortunate, suggesting, for example, a paradigm at the metalevel.

10. Masterman, "Nature of a Paradigm", p. 67.

11. A. Smith, *Powers of Mind* (New York: Ballantine, 1975), p. 19, connects Kuhn (see Thomas S. Kuhn, *The Structure of Scientific Revolutions*, 2nd ed. [Chicago: Chicago University Press, 1970]) with Conant (Kuhn was instructor for some of Conant's classes). Conant made heavy use of the notion of conceptual scheme in his account of science. The history of the recent use of "paradigm" in philosophical theory remains obscure. It has been suggested that Kuhn derived the term from Wittgenstein, who seems occasionally to have used it (or one of its German equivalents) in the relevant sense, for example: "What people are really after is something quite different. A certain paradigm hovers before their mind's eye, and they want to bring the calculus *into line with this paradigm*" (Thomas S. Khun, *Philosophical Remarks* [Oxford:

Blackwell, 1975], p. 346). The different more historical use of "paradigm" that features in the *paradigm case argument* also apparently derives from Wittgenstein and Moore.

The notion of "conceptual scheme" has of course several predecessors in European philosophy: *forms of understanding* (e.g. Kant): *forms of consciousness* (e.g. Marx): *forms of life* (in one sense, e.g. Wittgenstein).

12. Drengson, "Shifting Paradigms", p. 223; cf. also Masterman, "Nature of a Paradigm", p. 74, recalling the title of Black's work.

13. Masterman, "Nature of a Paradigm", pp. 66—67.

14. Ibid., p. 69.

15. Kuhn, *Scientific Revolutions*, p. 10.

16. Similar objections destroy familiar attempts to define paradigms in terms of features of the two features, for instance in terms of premises shared by a community of scientists. Also, paradigms are then ascribed to communities generally.

17. Kuhn, *Scientific Revolutions*, p. 10. The account indicated only applies to moderately well-developed cases; for in earlier stages there may be *no* laws, and little theory.

18. See the *Oxford English Dictionary* and, for the very diverse uses of the term in the sciences, L. Apostel, "Towards the Formal Study of Models in the Non-formal Sciences" in *Colloquium on the Concept and the Role of the Model in Mathematics and the Natural and Social Sciences*, ed. H. Freudenthal (Dordrecht: Reidel, 1961), pp. 1—37.

19. Masterman, "Nature of a Paradigm", p. 82.

20. Drengson, "Shifting Paradigms", p. 222.

21. Masterman, "Nature of a Paradigm", p. 82, says, surely mistakenly, that "philosophically speaking, a paradigm is an artefact which can be used as a puzzle-solving device. . . . " The account she suggests of a *crude* paradigm as, for example, literally a picture, differs from Kuhn's main suggestions, and seems to lead away from what is required for an account of scientific practice. Her comparison of paradigms with Black's *archetypes*, introduced in explaining models and metaphors, is nearer the target.

22. The point appears to be valid for almost all of the paradigm jargon in the social sciences, *once* preliminary rectification has been done. Often this preliminary work will be quite extensive, as G. Ritzer illustrates in *Sociology: A Multiple Paradigm Science* (Boston: Allyn and Bacon, 1975). Ritzer states that the following definition is the basis of his "entire book" on sociology:

A paradigm is a fundamental image of the subject matter within a science. It serves to define what should be studied, what questions should be asked, how they should be asked, and what rules should be followed in interpreting the answers obtained. The paradigm is the broadest unit of consensus within a science and serves to differentiate one scientific community (*or sub-community*) from another. It subsumes, defines, and interrelates the exemplars, theories, and attitudes and instruments that exist within it.

(Idem, *Towards An Integrated Social Paradigm* [Boston: Allyn and Bacon, 1981], p. 7.)

Yet two pages earlier he says of the first sentence, which carries the definition, that "this definition, simple and manageable" as it is, is of such a general nature that it would prove useless were we to apply it in any depth. "Therefore we must look for a more precise definition of this pivotal concept" (Ibid., p. 4). Evidently some attempt at consistency is required before the merry business of characterizing such an "image" in terms of a discipline-wide model is undertaken.

23. Finding a suitably general definition of *model* is rather like trying to find a

satisfactory definition of *market* (or often any definition at all) in mainstream economic texts uniformly advocating "market economies" – a possible needle in a haystack.

24. See, for example, Richard Routley "Universal Semantics", *Journal of Philosophical Logic* 4 (1975):327–56.

25. A universal model is representable as a structure M = ⟨T, K, D,∇, I⟩ where T, the base world, is in K, the set of worlds, D and ∇ are domain and subdomain functions from structure labels whose values are non-null sets, and I is an interpretation or valuation function. For full details see Routley, "Universal Semantics".

26. This claim has, like most large claims, been challenged, but not refuted.

27. As Vendler and Austin have taught us, a careful perusal of dictionaries can take us a long way in philosophical production. In this case the process was reversed. But for Smith's claim (which on the strength of some Webster's dictionaries such as the *New Collegiate,* I doubted) that "paradigm" meant "model", "according to the dictionary", I should not have persisted with further dictionaries: Webster's *New World* confirms Smith's claim. See Smith, *Powers of Mind.*

28. Forms of understanding, consciousness, ecological consciousness, etc., can also be accounted for as *assimilated* (or *internalized*) *models* of the same order of generality as cultural paradigms, presuppositions, and so on; the general theory of models goes some way to unifying important underlying models of much modern philosophy.

29. Imre Lakatos, "Falsification and the Methodology of Scientific Research Programmes", in *Criticism and Knowledge,* ed. Lakatos and Musgrave, pp. 91–196. The same applies to what Laudan said much earlier about "research traditions", the body of knowledge taken to be evolved according to its problem-solving capacity.

30. Lakatos, "Falsification", p. 132.

31. This can be sharpened through the formal notions of acceptance (⊢) and rejection (⊣).

32. Thus, for instance, talk of economic paradigms, political-science paradigms, sociological paradigms, where these are intended to apply in each case to the whole subject. There are also competing paradigms *within* these subjects.

33. "Perceiving", as "seeing", in the senses linked to "understanding". This carries no commitment to different ways of hearing, smelling, or seeing everyday medium-sized material objects.

34. Rodman, "Paradigm Change", p. 75.

35. See, e.g. Ophuls, *Politics of Scarcity,* pp. 222–23; W. R. Catton Jnr. and R. E. Dunlap, "New Ecological Paradigm for Post-exuberant Sociology", *American Behavioural Scientist* 24, no. 1 (1980):16.

36. Ibid., p. 34.

37. For example, Ophuls, *Politics of Scarcity,* p. 233; Paul Samuelson, *Economics,* 11th ed. (New York: McGraw Hill, 1980), p. 786.

38. As, for example, in the case of social psychology, R. C. Rosnow, *Paradigms in Transition* (New York: Oxford University Press, 1981).

39. Consider, for example, *public interest* and *safety* entries. The first presumably means that the economy is operated – somehow – in the public interest.

40. Even on a contemporary socialist view, the market forces, for example, would be regulated, and so on.

41. Rodman, "Paradigm Charge", pp. 49–78.

42. These are entirely *northern* continental paradigms. There are no comparable *southern* (continental) paradigms, for the southern continents are dominated

(economically and militarily and not independently thereof) ideologically by the north. Needed southern challenges to and overthrow of characteristically damaging northern paradigms (of political economy, and all so determined thereby at least) will be considered elsewhere.

43. Strictly, at this level there is a tradition of positions, with pure market and pure command economies at the extremes. Similarly at lower levels, there are often gradations, as the data compiled in table 3 helps to reveal. With paradigm contrasts, frequently a black or white, or few-colour, picture has been imposed on a full-colour situation.

44. By Rodman, "Paradigm Change", and others.

45. Daly, *Steady-State Economy*, p. 152 defines a steady-state economy by four (unsatisfactory) features:

| Constancy requirements | (1) | a constant human population |
| | (2) | a constant stock of artefacts |

| Value assumptions | (3) | the level at which (1) and (2) are to be constant are sustainable for a long future and sufficient for a good life |

| Minimization principle | (4) | the rate of throughput of matter-energy is reduced to lowest feasible level. |

These requirements would presumably bring much else in their train. In fact, requirement (3) appears difficult to satisfy without major political changes. But if only (1) and (2) were imposed, without (3), then severe distributional issues could arise along with probable political instability.

46. Drengson, "Shifting Paradigms", pp. 221–40.

47. It is tempting, and legitimate, to try to go deeper and to derive these differences and others from differences in underlying philosophical theory, in *this* case the "reference theory" versus the opposite. A beginning is made on such a project in Routley, *Exploring Meinong's Jungle and Beyond* (Canberra: Research School of Social Sciences, Australian National University, 1979).

48. A. Marsh, *Protest and Political Consciousness* (Beverly Hills: Sage, 1977), p. 106.

49. Such questions were asked at Simon Fraser University (SFU) by P. Hanson and S. Davis. I am indebted for comments to people at SFU and at McMaster University, where earlier and even rougher versions of this paper were presented, and also to John Rodman, Alan Drengson and Nicholas Griffin.

50. Marsh, *Protest and Political Consciousness*, p. 174. What appears is that the "personal values battery is tapping the 'postmaterialist phenomenon' rather more successfully . . . than the batteries developed . . . for the public value domain" (Ibid., p. 183). In conclusion:

> Essentially, post-materialists in the public domain are politically cynical, moderately efficacious, moderately aware leftists whereas post-materialists in the personal value domain tend to be moderately *trusting*, highly efficacious, politically sophisticated, and are normally distributed along the left-right continuum, clustering a fraction left of centre (Ibid., p. 184).

51. See especially, ibid., p. 176, table 7.1.

52. See table 3, p. 000; also see R. Inglehart, *The Silent Revolution: Changing Values and Political Styles among Western Publics* (Princeton, N.J.: Princeton University Press, 1977); and Marsh, *Protest and Political Consciousness*, pp. 165ff.

53. What has, however, been much disputed, and with good reason, is Inglehart's reactionary and highly economist explanation of this "silent revolution".

54. Marsh, *Protest and Political Consciousness*, pp. 165ff.

55. Part of this thesis Drengson does not render assessible, as he offers no account of "post-industrial culture". But his apparent assumption that post-industrial culture can be correlated with person-planetary paradigm rests on a mistake. The idea of post-industrial society, advanced by Bell and others in the 1960's (and summarized, e.g., in S. P. Huntington, "Post-Industrial Politics: How Benign Will It Be?", *Comparative Politics* 6 [1973–74]: 163–91) diverges drastically from person-planetary ideals: post-industrial society and culture represents rather the next (awful) stage of development of the technocratic paradigm (as heralded by advocates of that paradigm).

56. These attitudes afford one important reason for elaborating environmental positions and paradigms: see Richard and Val Routley, "Human Chauvinism and Environmental Ethics"; and idem, "Social Theories, Self-Management and Environmental Politics", in *Environmental Philosophy*, ed. Don Mannison, Michael McRobbie and Richard Routley, (Canberra: Research School of Social Sciences, Australian National University, 1970), pp. 96–189 and 217–332. Moreover, insofar as philosophy should be keeping track of popular thought, it should presumably be articulating and elaborating new positions: here is part of the case for environmental philosophy. A further part is that it should also be *anticipating* new directions.

57. See, e.g., Richard and Val Routley, "Human Chauvinism"; idem, "Environmental Politics"; also, Routley, *Fight for the Forests*.

58. Cognitive-dissonance theory explains attempts to bring ideas and attitudes into line with behaviour through an alleged human drive to avoid inconsistency.

59. In the environmental case, also, attitudes expressed may differ from practices. This may be because the attitudes are not genuinely held or are only disingenuously held. But it may be because, though the attitudes are held, the structural framework of social life does not readily permit, or punishes, the expression of these attitudes in practice. Again the latter can happen in a range of ways, from nonavailability of choice, through social and political pressure, to explicit punishment.

60. Writers with Marxist sympathies have other labels, for instance, *monopoly capitalism* (oligopic capitalism would be more accurate), *paternalistic capitalism*, etc.

61. Many of the questions build in false dichotomies in rather the way this one does (e.g., public/private, capitalist/socialist, etc).

62. "Businessmen say that they want less government in business, but that is what they say and not what they want. They are always coming to a legislature seeking regulation. They wish to have the state use its licensing power to give them a competitive advantage over other businessmen, especially over those outside the state" (D. D. McKean, *Pressures on the Legislature of Jersey* [New York: Columbia University Press, 1938], p. 56).

63. Although this is an assumption of most texts on organization theory, perhaps they don't always: the connection does not appear to be analytic.

64. J.K. Galbraith, *The New Industrial State* (London: Hamish Hamilton, 1967).

65. D. K. Hart and W. H. Scott, *Organizational America* (Boston: Houghton and Mifflin, 1979).

66. This familiar issue is discussed, for example, in the final chapter of Ophuls, *Politics of Scarcity*. There are, for this sort of reason among others, severe, though not insuperable, difficulties for proposals to *rationally* induce paradigm shifts (at least as regards higher paradigms). Still the situation is not as desperate or inevitably irrational as Kuhn and others suggest. According to Kuhn, *Scientific Revolutions*, p. 44, the arguments for a new paradigm may be "immensely persuasive" but that they cannot "be made logically or even

probabilistically compelling" except to people who are prepared to "step into the circle" (i.e., adopt the paradigm). The reason is that "when paradigms enter, as they must, into a debate about paradigm choice, their role is necessarily circular. Each group uses its own paradigm to argue in that paradigm's defense." This neglects such strategies as those of isolating critical elements or parts of paradigms (submodels), as well as the extent to which rational argument characteristically falls short of affording logical compulsion. A familiar but flawed high redefinition of "rational argument" underlies the irrationality theme.

67. For example, why utilities are ceasing to invest in new nuclear plants is explained in part by environmental action against nuclear plants and the impact of that action.

68. This is irrespective of whether the world system eventually collapses in some way and to some extent or not.

69. Part of the fuller argument for this theme may be found in the final, eloquent chapter of R. G. Collingwood, *An Autobiography* (Oxford: Oxford University Press, 1951). This intrusion of Collingwood into the picture is by no means coincidental. For Collingwood's theory of (metaphysical) presuppositions has much in common with more recent work in terms of paradigms and theory-dependence. A further use to which the explication and explanation of paradigms through models might be put is to tidy up formally, and integrate into the fuller theory, Collingwood's theory of presupposition.

Further Reading

Books

Ackerman, Bruce. *Private Property and the Constitution.* New Haven: Yale University Press, 1977.

————,ed. *Economic Foundations of Property Law.* Boston: Little Brown, 1975.

————,et al, *The Uncertain Search for Environmental Quality.* New York: Free Press, 1974.

Adler, C.A. *Ecological Fantasies.* New York: Glen Eagle Press, 1973.

Aiken, William, and La Follette, Hugh, eds. *World Hunger and Moral Obligation.* Englewood Cliffs: Prentice-Hall, 1977.

Allsop, Frederick. *Ecological Morality.* London: Frederick Muller, 1972.

Arrow, Kenneth. *Social Choice and Individual Values.* New Haven: Yale University Press, 1963.

Arvill, Robert. *Man and Environment: Crisis and the Strategy of Choice.* Harmondsworth: Penguin, 1973.

Ashby, Eric. *Reconciling Man with the Environment.* Oxford: Oxford University Press, 1978.

Baden, John and Hardin, Garrett. *Managing the Commons.* San Francisco: W.H. Freeman, 1977.

Baier, Kurt and Rescher, Nicholas, eds. *Values and the Future: The Impact of Technology on American Values.* New York: Free Press, 1969.

Barbour, Ian G., ed. *Earth Might Be Fair: Reflections on Ethics, Religion and Ecology.* Englewood Cliffs: Prentice-Hall, 1972.

Barkley, P.W., and Seckler, D.W. *Economic Growth and Environmental Decay.* New York: Harcourt, Brace and Jovanovich, 1972.

Barry, Brian, and Sikora, Richard, eds. *Obligations to Future Generations.* Philadelphia: Temple University Press, 1978.

Bateson, Gregory. *Steps to an Ecology of Mind.* San Francisco: Chandler Publishing Co., 1972.

Baxter, William F. *People or Penguins: The Case for Optimal Pollution.* New York and London: Columbia University Press, 1974.

Becker, Lawrence. *Property Rights.* London: Routledge and Kegan Paul, 1977.

Berg, Peter, and Dasmann, Raymond. *Reinhabitating a Separate Country.* San Francisco: Planet Drum Foundation, 1978.

Birch, Charles. *Confronting the Future.* Harmondsworth: Penguin, 1975.

Black, John. *The Dominion of Man: The Search for Ecological Responsibility.* Chicago: Aldine Publishing Co., 1970; and Edinburgh: Edinburgh University Press, 1970.

Blackstone, William T., ed. *Philosophy and Environmental Crisis.* Athens, G.: University of Georgia Press, 1974.

Boksenbaum, Howard; Commoner, Barry; and Corr, M., eds. *Energy and Human Welfare – A Critical Analysis.* Vol. 3. New York: Macmillan, 1975.

Bolander, K., and Claus, G. *Ecology Sanity: A Critical Examination of Bad Science, Good Intentions and Premature Doomsday Announcements of the Ecology Lobby.* New York: David McKay, 1974.

Bookchin, Murray. *Post-Scarcity Anarchism.* Berkeley, Calif.: Ramparts Press, 1971.

Bosselman, Fred, and Callies, David. *The Quiet Revolution in Land Use Control.* Washington, D.C.: G.P.O. Council on Environmental Policy, 1971.

Bremner, Michael J. *The Political Economy of America's Environmental Dilemma.* Lexington, Mass.: Lexington Books. New York: Praeger Publishers, Inc., 1973.

Brown, Martin, ed. *The Social Responsibility of the Scientist.* New York: Free Press, 1971.

Bryant, R.W.G. *Land: Private Property, Public Control.* Montreal: Harvest House, 1972.

Buchanan, James, and Tullock, Gordon. *The Calculus of Consent: Logical Foundations of Constitutional Democracy.* Ann Arbour: Michigan University Press, 1962.

Burch, W.R. *Daydreams and Nightmares: A Sociological Essay on the American Environment.* New York: Harper and Row, 1971.

Cahn, Robert. *Footprints on the Planet: The Search for an Environmental Ethic.* New York: Universe Books, 1978.

Caldwell, Lynton K. *Environment: A Challenge for Modern Society.* Garden City, N.Y.: Natural History Press, 1970.

————,et al. *Citizens and the Environment: Case Studies in Popular Action.* Bloomington: Indiana University Press, 1976.

Callahan, Daniel J. *The Tyranny of Survival; and Other Pathologies of Civilized Life.* New York: Macmillan, 1973.

Callenbach, Ernest. *Ecotopia.* Berkeley, Calif.: Banyan Tree Books, 1975.

Caroll, Peter. *Puritanism and the Wilderness.* New York: Columbia University Press, 1969.

Carson, Rachel. *Silent Spring.* Boston: Houghton Mifflin, 1962.

Carter, David, and Stringer, Peter. *Environmental Interaction.* Guildford, England: Surrey University Press, 1975.

Clark, Elizabeth, and Van Horn, Andre J. *Risk-Benefit Analysis and Public Policy: A Bibliography.* Cambridge: Energy and Environmental Policy Center, Harvard University Press, 1976.

Clark, Kenneth. *Landscape into Art.* London: John Murray, 1949.

Clark, Stephen R.L. *The Moral Status of Animals.* Oxford: Clarendon Press, 1977.

Cobb, John B., Jr. *Is It Too Late? A Theology of Ecology.* Milwaukee: The Bruce Publishing Co., 1971.

Coe, Cigi, and DeMoll, Lane, eds. *Stepping Stones: Appropriate Technology and Beyond.* New York: Schocken, 1978.

Collier, John. *Indians of the Americas: The Long Hope.* New York: American Library of World Literature, 1947.

———.*On the Gleaming Way.* Chicago: Swallow Press, 1962.

Commoner, Barry. *The Closing Circle.* London: Jonathan Cape, 1972.

———.*The Poverty of Power: Energy and the Economic Crisis.* New York: Knopf, 1976.

Connelly, Phillip, and Perlman, Robert. *The Politics of Scarcity: Resource Conflicts in International Relations.* Oxford: Oxford University Press, 1975.

Daly, Herman E., ed. *Towards a Steady-State Economy.* San Francisco: W.H. Freeman, 1973.

Disch, Robert, ed. *The Ecological Conscience.* Englewood Cliffs, N.J.: Prentice-Hall, 1970.

Disch, Robert, and Harney, R.T., eds. *The Dying Generations: Perspectives on the Environmental Crisis.* New York: Dell, 1971.

Dorfman, Robert, and Dorfman, Nancy S., eds. *Economics of the Environment.* New York: Norton, 1972.

Dubos, Rene. *The Wooing of Earth.* London: The Athlone Press, 1980.

Dubos, Rene, and Ward, B. *Only One Earth.* New York: Norton, 1972.

Duffey, Eric. *Conservation of Nature.* New York: Collins, 1970.

Ehrlich, Paul. *The Population Bomb.* New York: Ballantine, 1971.

Ehrlich, Paul, and Pirages, Dennis. *Ark II: Social Responses to Environment Imperatives.* San Francisco: Freeman, 1974.

Ehrlich, Paul; Ehrlich, Anne; and Holdren, John. *Ecoscience: Popula-*

tion, Resources, Environment. San Francisco: W.H. Freeman, 1977.

Fabos, Julius G., and Lawolmstead, Frederick, Jr. *Founder of American Landscape Architecture.* Amherst, Mass: Amherst Press, 1968.

Falk, Richard. *This Endangered Planet.* New York: Random House, 1971.

Feiverson, H.A., ed. *Boundaries of Analysis: An Inquiry into the Tocks Island Dam Controversy.* Cambridge, Mass.: Ballinger, 1976.

Ferkiss, Victor. *The Future of Technological Civilization.* New York: Braziller, 1974.

Flander, Susan L. *Thinking Like a Mountain: Aldo Leopold and the Evolution of an Ecological Attitude Toward Deer, Wolves and Forests.* Columbia, Miss.: University of Missouri Press, 1974.

Gabriel, Ralph. *American Values: Continuity and Change.* Westport, Conn.: Greenwood Press, 1974.

Galbraith, John Kenneth. *The Affluent Society.* London: Hamilton, 1958.

Glacken, C.J. *Traces on the Rhodian Shore.* Berkeley and Los Angeles: University of California Press, 1967.

Godlovitch, Stanley; Godlovitch, Rosalind; and Harris, John, eds. *Animal, Men and Morals.* London: Gollancz, 1971; New York: Taplinger Publishing Co., 1972.

Goodpaster, K.E., and Sayre, K.M., eds. *Ethics and Problems of the 21st Century.* Notre Dame: University of Notre Dame Press, 1979.

Graham, F. *Since Silent Spring.* Boston: Houghton-Mifflin, 1970.

Haefele, Edwin, T. *Representative Government and Environmental Management: Resources for the Future.* Baltimore: John Hopkins Press, 1973.

Hall, Gus. *Can We Survive Under Capitalism?* New York: International Publishers, 1974.

Hardesty, J., and Johnson, W., eds. *Economic Growth Versus the Environment.* Belmont, Calif.: Wadsworth Publishing Company, 1971.

Hardin, Garrett. *Nature and Man's Fate.* New York: Holt, Rinehart and Winston, 1959.

————.*Population, Evolution and Birth Control: A Collage of Controversial Ideas.* San Francisco: W.H. Freeman, 1969.

————.*Exploring New Ethics for Survival.* New York: Viking, 1972.

————.*The Limits of Altruism.* Bloomington: Indiana University Press, 1977.

Harrison, Ruth. *Animal Machines.* London: Stuart, 1974.

Hays, Samuel P. *Conservation and the Gospel of Efficiency.* New York: Atheneum, 1969.

Helfrich, H.H., Jr. *Agenda for Survival.* New Haven: Yale University Press, 1970.

————.ed. *The Environmental Crisis.* New Haven: Yale University Press, 1970.

Hirsch, Fred. *Social Limits to Growth.* Cambridge, Mass.: Harvard University Press, 1976.

Holum, John R. *Topics and Terms in Environmental Problems.* New York: Wiley, 1977.

Hooker, C.A., and Van Hulst, R. *Institutions, Counter-Institutions and the Conceptual Framework of Energy Policy Making in Ontario.* Research Report to the Royal Commission on Electric Power Planning, Ontario, Canada, May 1977.

Hughes, J. Donald. *Ecology and Ancient Civilization.* Alberquerque: University of New Mexico Press, 1975.

Illich, Ivan. *Energy and Equity.* London: Calder and Boyars, 1974.

Jaffe, L.L., and Tribe, L.H., eds. *Environmental Protection.* Chicago: Bracton, 1971.

Jarrett, Henry, ed. *Environmental Quality in a Growing Economy.* Baltimore: John Hopkins University Press, 1966.

Kormondy, E.J. *Concepts of Ecology.* Englewood Cliffs, N.J.: Prentice-Hall, 1969.

Kozlousley, Daniel E., ed. *An Ecological and Evolutionary Ethic.* Englewood Cliffs, N.J.: Prentice-Hall, 1974.

Krutilla, J.V., and Fischer, A.C., eds. *The Economics of Natural Environment.* Baltimore: John Hopkins, 1975.

Leiss, William. *The Domination of Nature.* New York: Braziller, 1972.

Leopold, Aldo. *A Sand County Almanac.* New York: Oxford University Press, 1949.

Linzey, Andrew. *Animal Rights: A Christian Assessment of Man's Treatment of Animals.* London: S.C.M. Press, 1976.

Lovins, Amory, and Price, John. *Non-Nuclear Futures: Case for an Ethical Energy Strategy.* San Francisco: Friends of the Earth, 1975.

Lowrance, William. *Of Acceptable Risk.* Los Altos: Wm. Kaufmann, 1976.

Luce, R.D., and Raiffa, H. *Games and Decisions.* New York: Wiley, 1957.

McHarg, Ian. *Design with Nature.* New York: Natural History Press, 1969.

MacPherson, C.B., ed. *Property: Mainstream and Critical Positions.* Toronto: University of Toronto Press, 1978.

Malthus, Thomas Robert. *First Essay on Population.* 1798. Reprint. New York: Kelley, 1965.

Mannison, Don; McRobbie, Michael; and Routley, Richards, eds.

Environmental Philosophy. Canberra: Research School of Social Sciences, Australian National University, 1980.

Marsh, George Perkins. *Man and Nature*. 1864. Reprint edited by David Lowenthal. Cambridge: Harvard University Press, 1965.

Marx, Leo. *The Machine in the Garden: Technology and the Pastoral Ideal in America*. Oxford: Oxford University Press, 1964.

Meadows, Dennis. *Alternatives to Growth*. Cambridge: Ballinger, 1977.

Meadows, Donella H., et al. *The Limits to Growth*. New York: Universe, 1972.

Meeker, Joseph. *The Comedy of Survival: Studies in Literary Ecology*. New York: Scribners, 1974.

Mendlovitz, Saul H., ed. *On the Creation of a Just World Order: Preferred Worlds for the 1990s*. New York: Free Press, 1975.

Mesarovic, Mihajlo, and Pester, Eduard. *Mankind at the Turning Point*. New York: New American, 1976.

Mishan, E.J. *The Costs of Economic Growth*. London: Staples Press, 1967.

————.*Cost-Benefit Analysis*. New York: Praegar, 1971.

————.*Elements of Cost-Benefit Analysis*. London: Allen and Unwin, 1972.

————.*The Economic Growth Debate: An Assessment*. London: Allen and Unwin, 1977.

Moos, R.H., and Brownstein, R. *Environment and Utopia*. New York: Plenum, 1977.

Muir, John. *A Thousand Mile Walk to the Gulf*. Boston: Houghton Mifflin, 1916.

Naess, Arne. *Freedom, Emotion and Self-Subsistence*. Oslo: Universiteforlaget, 1975.

Nash, Roderick. *The American Environment: Readings in the History of Conservation*. New York: Addison-Wesley, 1968.

————.*Environment and Americans*. New York: Holt Rinehart and Winston, 1972.

————.*Wilderness and the Mind*. New Haven: Yale University Press, 1973.

Neuhans, Richard. *In Defense of People*. New York: Macmillan, 1971.

Nozick, Robert. *Anarchy, State and Utopia*. Oxford: Blackwell, 1974.

Odum, Eugene. *Fundamentals of Ecology*. Philadelphia: Saunders, 1971.

Olson, Mancur. *The Logic of Collective Action: Public Goods and the Theory of Groups*. New York: Schocken, 1968.

————.*The Logic of Collective Action*. New York: Schocken, 1971.

————.*The No Growth Society*. New York: Norton, 1973.

Ophuls, William. *Ecology and the Politics of Scarcity*. San Francisco: W.H. Freeman, 1977.

O'Riordan, T. *Environmentalism.* London: Pion, 1976.

Paddock, Paul, and Paddock, William. *Famine – 1975.* Boston: Little, Brown and Co., 1968.

Parsons, H.L. *Marx and Engles on Ecology.* Westport, Conn.: Greenwood Press, 1977.

Passmore, John. *Man's Responsibility for Nature.* London: Duckworth, 1974.

Petulla, Joseph M. *American Environmental History: The Exploitation and Conservation of Natural Resources.* San Francisco: Boyd and Fraser, 1977.

Pinchot, Gifford. *Breaking New Ground.* New York: Harcourt, Brace and World, 1947.

Pursell, Carroll, ed. *From Conservation to Ecology: The Development of Environmental Concern.* New York: T. Y. Crowell, 1973.

Purtill, R.L. *Thinking about Ethics.* Englewood Cliffs, N.J.: Prentice-Hall, 1976.

Quarles, John. *Cleaning Up America.* Boston: Houghton-Mifflin, 1976.

Rawls, John. *A Theory of Justice.* Oxford: Oxford University Press, 1972.

Regan, Tom, and Singer, Peter, eds. *Animals Rights and Human Obligations.* Englewood Cliffs, N.J.: Prentice-Hall, 1976.

Regenstein, Lewis. *The Politics of Extinction.* New York: Macmillan, 1975.

Reilly, William K., ed. *The Use of Land: A Citizen's Policy Guide to Urban Growth.* New York: Crowell, 1973.

Roelofs, R.T., et al. *Environment and Society.* Englewood Cliffs, N.J.: Prentice-Hall, 1974.

Roszak, Theodore. *Where the Wasteland Ends.* New York: Doubleday, 1972.

Routley, Richard, and Routley, Val. *The Fight for the Forests.* 3rd ed. Canberra: Research School of Social Sciences, Australian National University, 1975.

Ryder, Richard. *Victims of Science.* London: Davis-Poynter, 1975.

Salt, H.S. *Animal Rights.* New York: Macmillan, 1894.

Sax, Joseph. *Defending the Environment: A Strategy for Citizen Action.* New York: Knopf, 1971.

Schumacher, E.F. *Small is Beautiful: A Study of Economics as if People Mattered.* London: Abacus, 1974.

Scoby, Donald, ed. *Environmental Ethics: Studies of Man's Self Destruction.* Minneapolis: Burgess Publishing Co., 1971.

Sen, Amartya K. *Collective Choice and Social Welfare.* San Francisco: Holden Day, 1970.

Shepard, Paul. *Living Animals.* New York: Viking Press, 1978.

Shrader-Frechette, K.S. *Nuclear Power and Public Policy.* Dordrecht, Holland: D. Reidel Publishing Co., 1978.

Sikora, Richard, and Barry, Brian, eds. *Obligations to Future Generations.* Philadelphia: Temple University Press, 1978.

Singer, Peter. *Animal Liberation: A New Ethic for Our Treatment of Animals.* New York: Avon, 1975.

————.*Practical Ethics.* Cambridge: Cambridge University Press, 1979.

Skolimowski, Henryk. *Ecophilosophy: Designing New Tactics for Living.* London: M. Boyers, 1979.

Smith, James N., ed. *Environmental Quality and Social Justice.* Washington, D.C.: Conservation Foundation, 1974.

Stone, C.D. *Should Trees Have Standing? Toward Legal Rights for Natural Objects.* Los Altos, Calif.: William Kaufman, 1974.

Stretton, Hugh. *Capitalism, Socialism and the Environment.* Cambridge: Cambridge University Press, 1974.

Strong, Douglas. *The Conservationists.* Reading, Mass.: Addison-Wesley, 1971.

Tribe, L.H., et al. *When Values Conflict.* Cambridge, Mass.: Ballinger, 1976.

Udall, Stewart. *The Quiet Crisis.* New York: Holt, Rinehart and Winston, 1963.

————.*Agenda for Tomorrow.* New York: Harcourt, Brace and World, 1968.

Weisberg, B. *Beyond Repair: The Ecology of Capitalism.* Boston: Beacon Press, 1971.

White, Lynn. *Machina ex Deo.* Massachusetts: MIT Press, 1970.

Worster, Donald. *American Environmentalism: The Formative Period, 1860–1915.* New York: Wiley, 1973.

————.*Nature's Economy: The Roots of Ecology.* San Francisco: Sierra Club Books, 1977.

Articles

There are numerous articles in which the philosophical aspects of environmental issues are discussed. The journal *Environmental Ethics* is entirely given over to such articles. *Zygon* is another useful source of such material. In addition, there are a number of journals in which material relevant to environmental philosophy is introduced and discussed. The following journals are useful in this regard: *Alternatives, Bioscience, The Ecologist, Ecologist Quarterly, Environment, Environment and Behaviour, Environmental Affairs, Environmental Education, Environmental Psychology, International Journal of Environmental*

Studies, Main Currents in Modern Thought, Natural Resources Journal, Policy Sciences, Polity, Science.

Some philosophy journals have devoted special issues to problems relevant to environmental philosophy. *Ethics* 88 (Jan. 1978) contains articles on animal rights, as does *Inquiry* 22, nos. 1–2 (1979) and *Philosophy,* 53, no. 206 (1978). The double issue of *Inquiry* contains a helpful bibliography. *Soviet Studies in Philosophy* 12 (Fall 1973) contains articles on the relationship between humankind and the natural environment.

The articles which follow provide useful starting points for further research.

Becker, Ernest. "Toward the Merger of Animal and Human Studies". *Philosophy of Social Sciences* 4 (1974): 235–54.

Blackstone, William T. "On the Rights and Responsibilities Pertaining To Toxic Substances". *Southern Journal of Philosophy* 16 (1978): 589–603.

Braybrooke, David. "From Economics To Aesthetics: The Rectification of Preferences". *Nous* 8 (1974): 13–24.

Clements, D.C. "Stasis: The Unnatural Value". *Ethics* 86 (1976): 136–43.

Danner, P.L. "Affluence and the Moral Ecology". *Ethics* 81 (1971): 287–302.

David, William. "Man-Eating Aliens". *Journal of Value Inquiry* 10 (1976): 178–85.

Dorst, Jean. "Current Problems of the Biosphere". *Diogenes* (Fall 1974): 85–105.

Golding, M.P., and Golding, N.H. "Ethical and Value Issues in Population Limitation and Distribution". *Vanderbilt Law Review* 24 (1971): 495–523.

Goodpaster, Kenneth. "On Being Morally Considerable". *Journal of Philosophy* 75 (1978): 308–25.

Hartshorne, Charles. "Beyond Enlightened Self-Interest: A Metaphysics of Ethics". *Ethics* 84 (1974): 201–16.

Hughes, J.D. "Ecology in Ancient Greece". *Inquiry* 18 (1974). 115–25.

Kainz, H.P. "Philosophy and Ecology". *The New Scholasticism* 47 (1973): 516–19.

Ladd, John. "Morality and the Ideal of Rationality in Formal Organisations". *The Monist* 54 (1970): 488–516.

McGinn, Thomas. "Ecology and Ethics". *International Philosophical Quarterly* 14 (1974): 149–60.

Naess, Arne. "The Shallow and The Deep, Long-Range Ecology Movement: A Summary". *Inquiry* 16 (1973): 95–100.

————."Spinoza and Ecology". *Philosophia* 7 (1977): 45–54.

Narveson, Jan. "Moral Problems of Population". *The Monist* 57 (1973): 62–86.

Passmore, John. "The Treatment of Animals". *Journal of The History of Ideas* 36 (1975): 195–218.

Rodman, John. "The Liberation of Nature?" *Inquiry* 20 (1976): 83–131.

Rolston III, Holmes. "Is There an Ecological Ethic?" *Ethics* 85 (1975): 93–109.

Roszak, Theodore. "Ecology and Mysticism". *Humanist* 86 (May 1971).

Routley, Richard, and Routley, Val. "Nuclear Energy and Obligations To The Future". *Inquiry* 21 (1978): 133–79.

Routley, Val. "Critical Notice of *Man's Responsibility For Nature*". *Australasian Journal of Philosophy* 53 (1975): 171–85.

Sagoff, Mark. "On Preserving the Natural Environment". *Yale Law Journal* 84 (1974): 205–67.

Seddon, George. "The Rhetoric and Ethics of the Environmental Protest Movement". *Meanjin* 31 (Dec. 1972): 427–37.

Shields, Allan. "Wilderness, Its Meaning and Value". *Southern Journal of Philosophy* 11 (1973): 240–53.

Travis, J.L. "Progressivism and the Human Supremacy Argument". *Philosophical Forum* 3 (1972): 208–21.

Tribe, Lawrence. "Policy Science: Analysis or Ideology?" *Philosophy and Public Affairs* 2 (1972): 66–110.